面向新工科"十三五"规划教材

Java 实践指导教程

主　编　曹德胜
副主编　张　玮　李永飞　王养廷

清华大学出版社
北京交通大学出版社
·北京·

内 容 简 介

本书以高校目前普遍使用的 Java 教材为背景，针对 Java 编程的特点，精心策划，准确定位，概念清晰，深入浅出，通过一些经典例题来阐述 Java 知识。每章分多个实践，每个实践都是先给出该实践程序的运行结果，再重点分析，这样可以使读者更容易理解和掌握。"程序分析与注意事项"是本书的特色，每章后都有习题和对部分习题的提示，来加深对本章知识的理解与掌握。

本书可作为高等院校计算机及相关专业的 Java 语言上机实践课程的指导书，也可以作为有一定 Java 语言基础知识的自学者的指导参考书。

本书封面贴有清华大学出版社防伪标签，无标签者不得销售。

版权所有，侵权必究。侵权举报电话：010-62782989　13501256678　13801310933

图书在版编目（CIP）数据

Java 实践指导教程/曹德胜主编．—北京：北京交通大学出版社：清华大学出版社，2019.10
ISBN 978-7-5121-4071-4

Ⅰ．①J… Ⅱ．①曹… Ⅲ．①JAVA 语言-程序设计-教材　Ⅳ．①TP312.8

中国版本图书馆 CIP 数据核字（2019）第 208069 号

Java 实践指导教程

Java SHIJIAN ZHIDAO JIAOCHENG

责任编辑：谭文芳

出版发行：清 华 大 学 出 版 社　　邮编：100084　电话：010-62776969　http://www.tup.com.cn
　　　　　北京交通大学出版社　　邮编：100044　电话：010-51686414　http://www.bjtup.com.cn

印　刷　者：北京鑫海金澳胶印有限公司

经　　销：全国新华书店

开　　本：185 mm×260 mm　　印张：11.25　　字数：283 千字

版　　次：2019 年 10 月第 1 版　　2019 年 10 月第 1 次印刷

书　　号：ISBN 978-7-5121-4071-4/TP·881

印　　数：1～2 000 册　　定价：36.00 元

本书如有质量问题，请向北京交通大学出版社质监组反映。对您的意见和批评，我们表示欢迎和感谢。
投诉电话：010-51686043，51686008；传真：010-62225406；E-mail：press@bjtu.edu.cn。

前　言

由于 Java 在面向对象的程序设计及其在网络开发中的广泛应用，引起了广大学生、计算机应用开发者的学习兴趣。不少学校已把 Java 程序设计作为计算机专业和相关专业的必修课程，为后继课程的学习打下良好的基础。为此，我们编写了这本《Java 实践指导教程》，以提高广大学习 Java 爱好者的上机操作能力。

本书通俗易懂，用一些经典的例题来由浅入深地阐述 Java 知识，把一些容易出现错误的地方在"程序分析与注意事项"中加以说明，以引起广大 Java 爱好者的注意。每章又分多个实践，每个实践都是先给出该实践例题的运行结果，再列出相关的知识点，最后对程序的执行加以分析，对程序编写中容易出错的地方加以提示，这样可以使读者更容易理解和掌握程序。

本书共分 8 章，第 1 章为 Java 语言基础，包括数据类型、运算符和表达式、分支和循环结构及其数组和向量等基础知识；第 2 章为系统类的应用，包括输入输出类、数据类型类、字符串处理类、数学计算类和日期时间类等系统类；第 3 章为用户自定义类，包括基本类的定义、类的继承、类的多态和重载、类的修饰、包的应用及其接口技术等；第 4 章为绘图基础和通用组件，包括文字、颜色和字体的设置方式、用户自定义绘图、图像处理、标签、按钮、输入框、选择框、鼠标键盘事件；第 5 章为容器和布局，包括"Applet""Panel""Frame"等容器，"FlowLayout""Borderlayout""Gridlayout""Cardlayout"等布局，"Canvas"画板、菜单设计、通用对话框及其事件容器等内容；第 6 章介绍异常及多线程，包括异常的基础知识、异常的处理、用户自定义异常、多线程基础、多线程的互斥与同步等知识；第 7 章为文件和多媒体，包括文件的基础知识、顺序文件、随机文件、标准输入和输出的重定向、多媒体技术基础等；第 8 章为网络编程，包括网页的打开、获取本地远程计算机的名称、IP 地址和工作组、"Socket"客户端和服务器的连接及其数据传送、网络通信原理等网络知识。

本书由曹德胜统筹并撰写 1~4 章，张玮撰写 5~8 章，另外参加编写的人员还有王养廷、李永飞等，全书由曹德胜完成统稿和审校工作。

由于编者水平有限和时间仓促，加之计算机技术发展十分迅速，本丛书必然会有一些不足之处甚至会出现或多或少的错误，敬请广大读者批评指正，必要时，可发送电子邮件至 1770779812@qq.com。

编　者

2019.8

目 录

第1章 Java 语言基础 1
实践 1-1　Java 运行环境 1
实践 1-2　数据类型 3
实践 1-3　运算符 5
实践 1-4　流程控制结构 8
实践 1-5　数组和向量 14
练习题 18

第2章 面向对象的程序设计（Ⅰ）——系统类的应用 20
实践 2-1　输入输出类 20
实践 2-2　数据类型类及其转换 22
实践 2-3　字符串处理类 25
实践 2-4　数学类和日期时间类 28
练习题 30

第3章 面向对象的程序设计（Ⅱ）——用户自定义类 32
实践 3-1　类的定义 32
实践 3-2　类的继承 35
实践 3-3　类的修饰 37
实践 3-4　类的多态 41
实践 3-5　类的封装 43
实践 3-6　接口技术 45
练习题 49

第4章 图形用户界面设计（Ⅰ） 53
实践 4-1　JFrame 基础 53
实践 4-2　基本几何图形的绘制 55
实践 4-3　图像的显示 58
实践 4-4　事件处理基础 61
实践 4-5　常用界面组件和键盘事件 63
实践 4-6　鼠标事件 69
练习题 73

第5章 图形用户界面设计（Ⅱ） 75
实践 5-1　常用布局 75
实践 5-2　动作和菜单 79
实践 5-3　选项对话框和自定义对话框 85
实践 5-4　文件选择器和颜色选择器 93

I

  练习题 ··· 102
第6章 异常及多线程 ··· 103
  实践 6-1 异常的基础知识 ··· 103
  实践 6-2 自定义异常 ··· 106
  实践 6-3 多线程入门 ··· 109
  实践 6-4 多线程的同步 ·· 115
  实践 6-5 阻塞队列 ·· 122
  练习题 ··· 130
第7章 文件 ··· 131
  实践 7-1 文件基本操作 ·· 131
  实践 7-2 二进制文件和对象序列化 ··· 139
  实践 7-3 内存映射文件 ·· 146
  练习题 ··· 149
第8章 网络编程基础 ··· 151
  实践 8-1 Socket 与 ServerSocket 编程 ··· 151
  实践 8-2 利用 URL 访问 Web 服务器 ·· 160
  实践 8-3 利用 JavaMail 发送 E-mail ··· 166
  练习题 ··· 171

第 1 章 Java 语言基础

运算符、表达式和控制结构、数组是任何一种程序语言的基础,Java 语言也不例外。

在这一章中,将向读者介绍如何在 Java Development Kit 和 Eclipse 环境下编辑、编译、解释 Java 语言,并学习 Java 语言的基本数据类型、运算符和表达式,以及流程控制结构、数组向量的应用。

实践 1-1 Java 运行环境

1. 实践结果
本例运行结果如图 1-1 和图 1-2 所示。

```
F:\JavaExperimentGuide\BookCodes\chapter1>javac HelloJava.java

F:\JavaExperimentGuide\BookCodes\chapter1>java HelloJava
HelloJava
```

图 1-1 Java 控制台应用在 Console 中运行

图 1-2 Applet 在 Eclipse 中运行

2. 实践目的
本例介绍 Java 程序运行的环境,了解如何用 JDK 解释器编译和解释 Console 环境下 Java 程序,如何用 Eclipse 编辑器编辑、编译、解释 Eclipse 环境下的 Java 程序。

1) Java 程序的运行过程

编辑:Java 具有小型化、平台无关性的特点,因此其代码可在任何一种编辑器中编辑,只要保存时，将文件的扩展名改为".java"即可。但推荐读者用NotePad++编辑器或Eclipse编辑。

编译：程序编辑结束后，须将程序转换成二进制代码，使机器能够识别。此转换过程，在Java中称为编译。经编译后的Java程序，其文件的扩展名为".class"，此时就可通过解释器对其解释。编译可用JDK的javac编译，也可用Eclipse直接编译。

解释：经编译后的Java程序，即class文件，还不能运行，须通过解释器对程序进行解释。解释可通过JDK的java解释（Console环境下）或javaAppletViwer解释（Eclipse环境下的Applet小程序），还可用Eclipse解释（Eclipse环境下）。

2) Eclipse编辑器

Eclipse编辑器是Windows环境下的Java编辑器，可以编辑Java程序，且在编辑时，会有代码提示，因此是编辑Java程序的好助手。其编辑环境如图1-3所示。

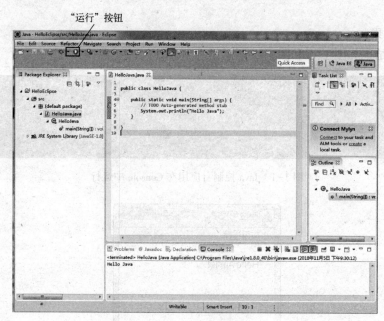

图1-3 Eclipse界面

3. 程序源代码

程序1的源代码如下：

```
public class HelloJava{
    public static void main(String args[]){
        System.out.println("HelloJava");
    }
}
```

程序2的源代码如下：

```
import java.applet.Applet;
public class HelloApplet extends Applet{
```

```
//Applet 小程序,用 JavaAppletViwer 解释器解释
public void paint(java.awt.Graphics g){
    g.drawString("Hello Applet",20,30);     //向浏览器中绘制文字
  }
}
```

4. 程序分析与注意事项

程序 1 是在 Console 环境下编写的一个 Java 小程序。其运行结果如图 1-1 所示。编完此程序后,需用 JDK 解释器编辑、解释程序。过程如下。

① 进入 Console 环境,将目录切换到 Java 程序存储的位置。

② 用户安装完成 JDK(在 https://www.oracle.com/technetwork/java/javase/overview/index.html 网站下载)。

③ 接着输入"javac Java 程序名"编辑程序。编辑完成后,生成 class 文件。需要注意的是:输入 Java 程序名时,一定要将文件的扩展名(.java)一起输入,否则将不能编译。

④ 最后输入 java CLASS 文件名。如果解释成功,将会在屏幕中显示相应的信息。需要注意的是:输入 CLASS 文件名时,不能将文件的扩展名(.class)一起输入,否则将不能解释。

程序 2 是在 Eclipse 环境下编写的 Java Applet 程序。其功能和程序 1 相似。其运行结果如图 1-2 所示。编完此程序后,需用 Eclipse 编辑和解释此程序。其过程很简单,只要单击工具栏的"运行"按钮即可,如图 1-3 所示。

实践 1-2　数据类型

1. 实践结果

本例运行结果如图 1-4 所示。

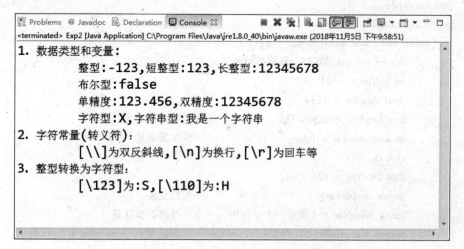

图 1-4　实践 1-2 在控制台的输出

2. 实践目的

本例介绍 Java 编程的基本结构。Java 定义了 8 种基本的数据类型。

1) 整型

整型可分为字节型（byte）、短整型（short）、整型（int）和长整型（long）4 种。

int 类型：给定系统的基本类型，用 4 个字节的存储单元存放。

long 类型：要表示一个数为长整型，需要在这个数后面加字母"L"或"l"。

2) 实型

实型数据分单精度（float）和双精度（double）两种类型。

float 类型：占用 4 个字节内存，有效数字为 7 位。需要在这个数后面加字母"F"或"f"。

double 类型：占用 8 个字节内存，有效数字为 15 位。

3) 字符型

字符常量是无符号的常量，占用 2 个字节内存。

4) 布尔型

布尔常量有两个值：true 和 false，分别代表布尔逻辑中的"真"和"假"。在 Java 中，布尔常量不能转换成任何数据类型，true 常量不等于 1，而 false 常量也不等于 0。这两个值只能赋给声明为 Boolean 类型的常量，或者用于布尔运算表达式中。

5) 字符串常量

一个字符串常量是用双引号括起来的 0 个或多个字符组成的序列。下面列举的都是合法的字符串常量。

""：为空串。

"Hello Java\n"：为一系列字符构成的字符串，其中的"\n"是回车换行符。

3. 程序源代码

程序的源代码如下：

```java
package exp2;
public class Exp2 {
    public static void main(String[ ] args) {
        System.out.println("1. 数据类型和变量:");
        int intVar = -123;
        short shortInt = 123;
        long longVar = 12345678;          //整型变量
        boolean boolVar = false;          //布尔型变量
        char ch = 'X';                    //字符型变量
        float floatVar = 123.456f;
        double doubleVar;                 //实型变量
        String stringVar = "我是一个字符串";  //字符串类型变量
        System.out.println("\t整型:" + intVar + ",短整型:" + shortInt + ",长整型:" + longVar);
        System.out.println("\t布尔型:" + boolVar);
        System.out.println("\t单精度:" + floatVar + ",双精度:" + longVar);
        System.out.println("\t字符型:" + ch + ",字符串型:" + stringVar);
```

System.out.println("2、字符常量(转义符):");
//分别在屏幕上输出"\\"、"\n"、"\r"。
System.out.println("\t[\\\\]为双反斜线,[\\n]为换行,[\\r]为回车等");
System.out.println("3、整型转换为字符型:");
//分别在屏幕上输出"S"和"H"。
System.out.println("\t[\\123]为:\123,[\\110]为:\110");
     }
  }

4. 程序分析与注意事项

此部分包括程序解释及常见问题。

① //:以双斜线"//"开头的内容是 Java 语言的注释部分。Java 语言中有三种注释形式:"//"表示在行注释。对程序的作用作简短的说明;"/* */"表示一行或多行的注释。这种注释可用于多行注释且"/*"和"*/"之间的内容都是注释内容;"/** */"表示文档注释,文档注释放在说明之前,表示该段注释应包含在自动生成的任何文档中。在程序中增加注释,是为了增加程序的可读性。

② "public class Exp2"语句的作用是定义一个主类,使用关键字 class,其后面是类的名称 Exp2,public 表示这个类的访问特性是公共的。以下用一对大括号括起来是类体。类的定义将在第 3 章中详细介绍。

③ "public static void main(String[ ] args)"语句的作用是定义了一个主函数,作为程序运行的入口。

④ "System.out.println("1.数据类型和变量:");"语句的作用是:在屏幕输出文字,这里调用 java.lang 包中的 System 类,而 System.out 又是 java.lang 包中 OutputSystem 类的对象,方法 println( )的作用是把圆括号内的字符串在屏幕输出且回车换行。

⑤ 在书写程序的过程中应注意,每个语句的结束都要加上";",否则在编译程序中会出现错误。出错信息为:"**Expected ';'**"。意思为"缺少';'"。

⑥ 在编译程序中,应特别注意几个常见的错误。例如:在编译过程中,应特别注意文件名的大小写。在本例中,如果将"Exp2.java"写成"exp2.java",则会在编译过程中出现错误,出错信息为:"**Can't read :Exp2.java**"。意思为"不能读取'Exp2'文件,即文件不存在。"

实践 1-3 运算符

1. 实践结果

本例运行结果如图 1-5 所示。

2. 实践目的

本例介绍 Java 运算符的应用,包括赋值运算符、类型转换、双目运算符、单目运算符、关系运算符、逻辑运算符、移位运算符、三目运算符。熟悉 Java 语言的运算符和表达式的

```
Console ☒
<terminated> Exp3 [Java Application] C:\Program Files\Java\jre1.8.0_40\bin\javaw.exe (2018年11月5日 下午10:08:17)
1. 赋值运算后，各变量的值为：one=101,two=50 ,three=0  ,four=0,bl=true
       表达式为：one=101,two=50,three=0,four=0,bl=true
2. 类型转换后，各变量的值为：one=101,two=50 ,three=500,four=0,bl=true
       表达式为：three=(int)ft;
3. 双目运算后，各变量的值为：one=101,two=50 ,three=500,four=2,bl=true
       表达式为：four=one/two
4. 单目运算后，各变量的值为：one=0  ,two=500,three=500,four=2,bl=true
       表达式为：two=++three
5. 关系运算后，各变量的值为：one=0  ,two=500,three=500,four=2,bl=false
       表达式为：bl=(boolean)(two<=one)
6. 逻辑运算后，各变量的值为：one=0  ,two=500,three=500,four=2,bl=true
       表达式为：bl=two==three||four>=one
7. 移位运算后，各变量的值为：one=0  ,two=500,three=125,four=2,bl=true
       表达式为：three=two>>four
8. 三目运算后，各变量的值为：one=0  ,two=500,three=125,four=2,bl=false
       表达式为：bl=(one!=100)&&two!=three?!(four>one):bl
```

图 1-5 实践 1-3 在控制台的输出

应用、各运算符之间的相互联系及其各运算符的优先级。

① 强制类型转换：在一个变量和表达式前添加一个类型符，可将其他类型的变量或表达式转换为此类型。其格式为：

(类型符)变量或表达式

② 双目运算符：加号（+）、减号（-）、乘号（*）、除号（/）、取余（%）。

③ 单目运算符：自加（++）、自减（--）、求反（-）。

④ 关系运算符：大于（>）、等于（==）、小于（<）、大于等于（>=）、小于等于（<=）、不等于（!=）。

⑤ 逻辑运算符：假与运算（&&）、真或运算（||）、与运算（&）、或运算（|）、非运算（!）、异或运算（^）。其中"假与运算"与"与运算"的区别为：运算中，只要表达式中的一个部分为假，则运算就结束的与运算称为"假与运算"，否则为"与运算"。

⑥ 移位运算符：位反（~）、右移（>>）、左移（<<）、无符号右移（>>>）。位移运算的实质是将一个数乘除 2 的 "n" 次方而已。"n" 为移位的位数。

⑦ 三目运算符：格式为"表达式?真值:假值"。即当表达式为真时，结果为"真值"，否则为"假值"。

⑧ 赋值运算符：将程序中要存储、传递的数据赋给一个变量或对象的运算叫做赋值运算。其格式为"变量或对象=表达式"，赋值运算符包括等号（=）、加号赋值（+=）、减号赋值（-=）、乘号赋值（*=）、除号赋值（/=）、取余赋值（%=）、与赋值（&=）、或赋值（|=）、异或赋值（^=）、右移赋值（>>=）、左移赋值（<<=）、无符号右移赋值（>>>=）。

⑨ 运算符的优先级：点号、方括号、括号→单目运算→双目运算→移位运算→逻辑运算→三目运算→赋值运算。

3. 程序源代码

程序的源代码如下：

```java
package exp3;

public class Exp3 {
    public static void main(String[] args) {
        int one = 101, two = 50, three = 0, four = 0;  //赋值运算符，初始化数据
        boolean bl = true;
        float ft = 500.12345f;
        System.out.println("1. 赋值运算后，各变量的值为：one=" + one + ",two=" + two
            + ",three=" + three + ",four=" + four + ",bl=" + bl);
        System.out.println("\t 表达式为：one=101,two=50,three=0,four=0,bl=true");
        three = (int) ft;
        System.out.println("2. 类型转换后，各变量的值为：one=" + one + ",two=" + two
            + ",three=" + three + ",four=" + four + ",bl=" + bl);
        System.out.println("\t 表达式为：three=(int)ft;");
        four = one / two;                //双目运算符，对 one 和 two 进行取整运算
        System.out.println("3. 双目运算后，各变量的值为：one=" + one + ",two=" + two
            + ",three=" + three + ",four=" + four + ",bl=" + bl);
        System.out.println("\t 表达式为：four=one/two");
        one = two % four;                //双目运算符，对 two 和 four 进行取余运算
        three--;                         //单目运算符，对 three 进行自减运算（先赋值再自减）
        two = ++three;                   //单目运算符，对 three 进行自加运算（先自加再赋值）
        System.out.println("4. 单目运算后，各变量的值为：one=" + one + ",two=" + two
            + ",three=" + three + ",four=" + four + ",bl=" + bl);
        System.out.println("\t 表达式为：two=++three");
        bl = (boolean)(two <= one);      //关系运算符，对 one 和 two 变量进行比较大小
        System.out.println("5. 关系运算后，各变量的值为：one=" + one + ",two=" + two
            + ",three=" + three + ",four=" + four + ",bl=" + bl);
        System.out.println("\t 表达式为：bl=(boolean)(two<=one)");
        bl = two == three || four >= one;//逻辑运算符，对 one 和 two 变量进行或运算
        System.out.println("6. 逻辑运算后，各变量的值为：one=" + one + ",two=" + two
            + ",three=" + three + ",four=" + four + ",bl=" + bl);
        System.out.println("\t 表达式为：bl=two==three||four>=one");
        three = two >> four;             //移位运算符，对将 two 向右移 four 位
        System.out.println("7. 移位运算后，各变量的值为：one=" + one + ",two=" + two
            + ",three=" + three + ",four=" + four + ",bl=" + bl);
        System.out.println("\t 表达式为：three=two>>four");
```

```
        bl = (one != 100) && two ! = three ? ! (four > one) : bl;//三目运算符
        System. out. println("8. 三目运算后,各变量的值为:one=" + one + "    ,two=" + two
                  + ",three=" + three + ",four=" + four + ",bl=" + bl);
        System. out. println("\t表达式为:bl=(one! =100)&&two! =three?! (four>one):bl");
    }
}
```

4. 程序分析与注意事项

此部分包括程序解释及常见问题。

① "int one=101,two=50,three=0,four=0;" 语句的作用是:定义了四个变量,并用赋值语句为每个变量赋值。这就是赋值运算符的应用。

② "three=(int)ft;" 语句的作用是:将浮点类型的变量转换为整型变量。要将一个类型转换为另一个类型,只要在变量前加类型说明符即可。

③ "four=one/two;" 语句是一个双目运算符运算的表达式。

④ "two=++three;" 语句的作用是:将变量 three 的值先赋给变量 two,再将变量 three 的值加1。与 "two=three++;" 语句的区别在于:"two=++three;" 语句是将变量 three 的值先加1,再将此时 three 变量的值赋给变量 two,假设 three 的初始值为1,运行 "two=++three;" 语句后,变量 two 的值为1,变量 three 的值为2;而运行 "two=three++;" 语句后,变量 two 的值为2,变量 three 的值为2。

⑤ "bl=(boolean)(two<=one);" 语句是一个关系运算符运算的表达式,其结果只有 "true" 或 "false" 两种。

⑥ "bl=two==three||four>=one;" 语句是一个逻辑运算符运算的表达式,其结果也为 true 或 false 两种。但逻辑运算符两边的表达式的值必须为 boolean 类型的值,即为关系运算符或逻辑运算符的表达式。

⑦ "three=two>>four;" 语句是一个位移运算符运算的表达式,其运算过程是:将变量 two 转换为二进制数,再将其向右移 four 位。需要注意的是 four 须为整数。移动后相当于将变量 two 的值除以2的 four 次方。例如:十进制数10,转换位二进制数为110,运行 "10>>1" 语句后二进制值为11,即十进制的值为5。

⑧ "bl=(one! =100)&&two! =three?! (four>one):bl;" 语句为一个三目运算符表达式。其功能是:如果表达式 "(one! =100)&&two! =three" 的值为真,则变量 "bl" 的值为 "!(four>one)" 表达式的值;否则变量 "bl" 的值为变量 "bl" 的值。

实践 1-4 流程控制结构

1. 实践结果

本例运行结果如图 1-6 所示。

2. 实践目的

本例介绍流程控制结构的应用,需要熟悉控制语句的用法,包括顺序,分支,循环结构;熟悉 break 和 continue 关键字的应用。

```
                        2019年日历
     日 一 二 三 四 五 六    日 一 二 三 四 五 六    日 一 二 三 四 五 六
           01 02 03 04 05                   01 02                   01 02
     06 07 08 09 10 11 12    03 04 05 06 07 08 09    03 04 05 06 07 08 09
     13 14 15 16 17 18 19    10 11 12 13 14 15 16    10 11 12 13 14 15 16
     20 21 22 23 24 25 26    17 18 19 20 21 22 23    17 18 19 20 21 22 23
     27 28 29 30 31          24 25 26 27 28          24 25 26 27 28 29 30
                                                     31

        01 02 03 04 05 06             01 02 03 04                      01
     07 08 09 10 11 12 13    05 06 07 08 09 10 11    02 03 04 05 06 07 08
     14 15 16 17 18 19 20    12 13 14 15 16 17 18    09 10 11 12 13 14 15
     21 22 23 24 25 26 27    19 20 21 22 23 24 25    16 17 18 19 20 21 22
     28 29 30                26 27 28 29 30 31       23 24 25 26 27 28 29
                                                     30

        01 02 03 04 05 06                01 02 03    01 02 03 04 05 06 07
     07 08 09 10 11 12 13    04 05 06 07 08 09 10    08 09 10 11 12 13 14
     14 15 16 17 18 19 20    11 12 13 14 15 16 17    15 16 17 18 19 20 21
     21 22 23 24 25 26 27    18 19 20 21 22 23 24    22 23 24 25 26 27 28
     28 29 30 31             25 26 27 28 29 30 31    29 30

           01 02 03 04 05             01 02          01 02 03 04 05 06 07
     06 07 08 09 10 11 12    03 04 05 06 07 08 09    08 09 10 11 12 13 14
     13 14 15 16 17 18 19    10 11 12 13 14 15 16    15 16 17 18 19 20 21
     20 21 22 23 24 25 26    17 18 19 20 21 22 23    22 23 24 25 26 27 28
     27 28 29 30 31          24 25 26 27 28 29 30    29 30 31
```

图 1-6 实践 1-4 在控制台的输出

1）分支结构

（1）if 语句

"if 语句结构"有以下格式：

  if(表达式) 语句 1;[else 语句 2;]
  if(表达式)
  {
    if(表达式){语句 1;}
    [else{语句 2;}]
  }
  [else
  {
    if(表达式){ 语句 3;}
    [else{语句 4;}]
  }]

语句中的表达式必须为关系运算符或逻辑运算符的表达式。如果表达式的值为真，则运行 else 语句前的语句，否则运行 else 语句后的语句。

（2）switch case 语句结构

switch case 语句的格式为：

```
switch(表达式)
{
  case 常量 1:语句块 1;[break;]
  case 常量 2:语句块 1;[break;]
  …
  case 常量 n:语句块 1;[break;]
  default:语句块 n+1;
}
```

此结构将根据表达式的值转到与其匹配的常量值去执行 case 语句，是一种多分支的结构。

2) 循环结构

（1）for 循环语句结构

for 循环语句的格式为：

　　for(表达式 1;表达式 2;表达式 3){循环体;}

其中，表达式 1 为初始化循环，它用来初始化循环的控制变量；表达式 2 为循环的终止条件，其值必须为逻辑类型，如果值为假，则退出循环，否则继续运行循环体；表达式 3 为循环控制变量的改变方式，只有通过改变循环变量才能使得循环遵循一定的规律，使其不会产生死循环。

（2）while 语句结构

　　while(表达式){循环体;}

其中，只要表达式的值为真就运行循环体，否则退出循环。

（3）do…while 语句结构

do…while 的格式为：

　　do{循环体;}while(表达式);

其中，程序先运行循环体一次，再判断表达式的值，如果为真，则继续执行循环体，否则退出循环。

3) break 关键字

当循环结构的程序运行时，遇到 break 关键字将强制退出当前循环。

4) continue 关键字

当循环结构的程序运行时，遇到 continue 关键字将不执行此关键字以下的循环体语句，而是继续执行下一次循环。

3. 程序源代码

程序的源代码如下：

```
package exp4;

public class Exp4 {
```

```java
public static void main(String[] args) {
    String Weeks = "   日   一   二   三   四   五   六   ";
    int i, j, day1 = 1, day2 = 1, day3 = 1, Ospa3 = 1;
    int spa1 = 0, spa2 = 0, spa3 = 0, cou1 = 0, cou2 = 0, cou3 = 0;
    //改变"year"的值(大于1900年),就可列出当年的日历表
    int year = 2019;
    //判断当年是否是闰年
    boolean leap = year % 4 == 0 && year % 100 ! = 0 || year % 400 == 0;
    //提起1900年后,"year"年1月1日的星期
    for (i = 1900; i < year; i++)
        if (i % 4 == 0 && i % 100 ! = 0 || i % 400 == 0)
            Ospa3 += 2;
        else
            Ospa3++;
    Ospa3 %= 7;

    System.out.println("\t\t\t\t"+year+"年日历");
    //在屏幕输出星期的对照
    for (i = 1; i <= 3; i++)
        System.out.print(Weeks);
    System.out.println("");
    for (j = 1; j <= 28; j++) {
        if ((j - 1) % 7 == 0) {
            day1 = 1;
            day2 = 1;
            day3 = 1;
            //分别设置一年中每个月1日的星期和当月的总天数
            switch (j / 7) {
                case 0:
                    spa1 = (Ospa3 + cou3) % 7;
                    cou1 = 31;                  //1月
                    spa2 = (spa1 + cou1) % 7;
                    cou2 = leap ? 29 : 28;      //2月
                    spa3 = (spa2 + cou2) % 7;
                    cou3 = 31;                  //3月
                    Ospa3 = spa3;
                    break;
                case 1:
                    spa1 = (Ospa3 + cou3) % 7;
                    cou1 = 30;                  //4月
                    spa2 = (spa1 + cou1) % 7;
                    cou2 = 31;                  //5月
                    spa3 = (spa2 + cou2) % 7;
```

```java
                    cou3 = 30;                    //6月
                    Ospa3 = spa3;
                    break;
                case 2:
                    spa1 = (Ospa3 + cou3) % 7;
                    cou1 = 31;                    //7月
                    spa2 = (spa1 + cou1) % 7;
                    cou2 = 31;                    //8月
                    spa3 = (spa2 + cou2) % 7;
                    cou3 = 30;                    //9月
                    Ospa3 = spa3;
                    break;
                case 3:
                    spa1 = (Ospa3 + cou3) % 7;
                    cou1 = 31;                    //10月
                    spa2 = (spa1 + cou1) % 7;
                    cou2 = 30;                    //11月
                    spa3 = (spa2 + cou2) % 7;
                    cou3 = 31;                    //12月
                    break;
            }
        }
        for (i = 1; i <= 21; i++) {
            //在屏幕上一行一行地输出日历
            if ((i - 1) / 7 == 0)
                System.out
                        .print(spa1-- <= 0 && day1 <= cou1 ? day1 < 10 ? " 0"
                                + day1++
                                : " " + day1++
                                : "    ");
            else if ((i - 1) / 7 == 1)
                System.out
                        .print(spa2-- <= 0 && day2 <= cou2 ? day2 < 10 ? " 0"
                                + day2++
                                : " " + day2++
                                : "    ");
            else
                System.out
                        .print(spa3-- <= 0 && day3 <= cou3 ? day3 < 10 ? " 0"
                                + day3++
                                : " " + day3++
                                : "    ");
            if (i % 7 == 0)
```

```
                    System.out.print("  ");
                }
                //换行
                System.out.println("");
            }
        }
    }
```

4. 程序分析与注意事项

此部分包括程序解释及常见问题。

① "boolean leap=year%4==0&&year%100!=0||year%400==0;"语句的作用是判断当年是否是闰年，其判断依据是："四年一闰，百年不闰，四百年又闰。"

② 以下程序段：

```
for(i=1900;i<year;i++)
    if(i%4==0&&i%100!=0||i%400==0) Ospa3+=2;
    else Ospa3++;
Ospa3%=7;
```

的作用是：因一年的总天数为 365 天（闰年为 366 天），则将其除以 7 得余数为 1（闰年为 2）。即下一年 1 月 1 日的星期为上一年 1 月 1 日的星期加 1（闰年加 2）。例 1998 年 1 月 1 日是星期四，则 1999 年的 1 月 1 日就为星期五，而 2000 年的 1 月 1 日为星期六（因为闰年）。表明星期是以 7 天为一个周期的。

"for(i=1900;i<year;i++)"语句是一个循环语句，作用是使变量"i"从 1900 开始循环；循环终止条件为小于变量"year"；循环变量的改变方式为自加 1。

"if(i%4==0&&i%100!=0||i%400==0) Ospa3+=2;else Ospa3++;"语句为一个分支结构，当"i%4==0&&i%100!=0||i%400==0"表达式的值为真时，就运行"Ospa3+=2"语句，否则运行"Ospa3++"语句。

③ 以下程序段：

```
switch(j/7)
{
    case 0:
    ……
```

的作用是：同上可得，下一个月 1 日的星期应为上一个月 1 日的星期加上一个月的总天数的和除以 7 的余数。例 2001 年 1 月 1 日为星期一，则 2 月 1 日应为(1+31)/7=4，即为星期四。"switch(j/7){ }"语句的作用是：将 j/7 的值与 case 后的常量值进行比较。如果与其中的一个 case 相匹配，则程序将转向此 case 的程序块中运行。需要注意的是，case 程序块的 break 关键字是使其退出此 case 语句块，如果没加此 break 关键字，则将向下一个 case 程序块继续运行。

④ "System. out. print(spa1--<=0&&day1<=cou1？day1<10?" 0"+day1++:" "+day1++:"   ")；"语句的作用是：要使得每月1日与星期相对应地显示在屏幕上，就得在开始前加若干个空格。空格数由此月1日的星期决定。此外，每个月的天数不能超过其总天数，超过部分也应用空格代替。所有的数字都是以两位数的方式显示，须在一位的数字前加0。

其运行过程为：如果"spa1--<=0&&day1<=cou1"表达式的值为真，就运行"day1<10?" 0"+day1++:" "+day1++"语句，此语句又是一个三目运算。如果"day1<10"表达式为真，运行"0"+day1++"表达式，否则运行" "+day1++"表达式。最后运行"   "。

实践1-5 数组和向量

1. 实践结果

本例运行结果如图1-7所示。

```
_____Fibonacci_____
1          1          2          3          5          8          13         21
34         55         89         144        233        377        610        987
1597       2584       4181       6765       10946      17711      28657      46368
75025      121393     196418     317811     514229     832040     1346269    2178309
_____Triangle of Yanghui_____
1
1    1
1    2    1
1    3    3    1
1    4    6    4    1
1    5    10   10   5    1
1    6    15   20   15   6    1
1    7    21   35   35   21   7    1
1    8    28   56   70   56   28   8    1
1    9    36   84   126  126  84   36   9    1
_____Deal with the Vector_____
          0 1 2 3 4 5 6 7 8 9 0 1 2 3 4 5 6 7 8 9
Init    Data:Love Java!
Insert  Data:I Love Java!
Search  Data:The Element "Ja" At:7
Modify  Data:I Love java!
Remove  Data:I Love java
```

图1-7 实践1-5在控制台的输出

2. 实践目的

本例介绍数组的定义、创建、赋值、访问的方式和向量的基本操作，包括添加、删除、修改、查找。熟悉数组的定义、初始化、赋值及其应用，向量的创建、访问的方式。

（1）数组

数组并不是一种数据类型，而是一组相同类型的变量的集合。创建一个数组后，系统将分配一段连续的空间供数组使用。

数组定义的格式为：

　　　　类型说明符 数组名[]；

或

　　类型说明符[ ] 数组名

数组创建的格式为：

　　数组名=new 类型说明符[数组元素的个数]；

或

　　数组名={初值表列}；

综上所述，数组可用一步的形式进行定义、创建、赋初值。格式为：

　　类型说明符 数组名[ ]={初值表列}；或类型说明符[ ] 数组名={初值表列}；

（2）向量

向量是指创建时先给对象一定的空间，当此空间不足时，系统将根据向量增量相应地增加内存空间，使得内存空间可被充分利用。其与数组的区别在于：数组创建时，分配了一段固定的空间，此空间不可改变。而向量的内存空间可以动态改变，定义向量的格式为：

　　Vector 向量名=new Vector(初始大小,向量增量)；

3. 程序源代码

程序的源代码如下：

```
package exp5;

import java.util.Vector;

public class Exp5 {

    public static void main(String[] args) {
        //TODO Auto-generated method stub
        int i;
        int Fibonacci[];                        //定义了一个一维数组
        Fibonacci=new int[32];                  //创建了一个一维数组
        Fibonacci[0]=1;Fibonacci[1]=1;          //给数组赋值
        //根据"Fibonacci"数列的算法列数据
        System.out.println("\t\t_____ Fibonacci _____");
        for(i=2;i<32;i++)
            Fibonacci[i]=Fibonacci[i-1]+Fibonacci[i-2];
            //访问数组的元素
        for(i=0;i<32;i++)
        {
            System.out.print(Fibonacci[i]+"\t");
            if((i+1)%8==0) System.out.println("");
```

```
        }
        int[ ][ ] TriangleYH;                          //定义了一个二维数组
        TriangleYH=new int[10][10];                    //创建了一个二维数组
            //根据杨辉三角的算法列出数据
        System.out.println("\t\t_____ Triangle of Yanghui _____");
        for( i=0;i<10;i++)
            for( int j=0;j<=i;j++)
                if(i==j||j==0)
                    TriangleYH[i][j]=1;
                else
                    TriangleYH[i][j]=TriangleYH[i-1][j-1]+TriangleYH[i-1][j];
        //访问数组的元素
        for( i=0;i<10;i++)
        {
            for( int j=0;j<=i;j++)
                System.out.print(TriangleYH[i][j]+"\t");
            System.out.println("");
        }
        System.out.println("\t\t_____ Deal with the Vector _____");
            //创建一个向量,初始化空间为8,增量为4
        Vector<String> MyVT=new Vector<String>(8,4);//出于类型安全的因素,建议使用泛型
        String str[ ]={"L","o","v","e"," ","Ja","va","!"};
        System.out.print("                ");
        for( i=0;i<20;i++)
            System.out.print(i%10+" ");
        System.out.println("");
        System.out.print("Init    Data:");
        for( i=0;i<8;i++)
        {
            MyVT.addElement(str[i]);                   //向向量中添加元素
            System.out.print(MyVT.elementAt(i));       //访问向量中的元素
        }
        System.out.println("");
        System.out.print("Insert Data:");
        /*继续向向量中添加元素,因此时向量中的元素个数超过初始化个数,
          向量的个数将按增量增加,则其元素的总个数为8+2=10个*/
        MyVT.insertElementAt("I",0);
        MyVT.insertElementAt(" ",1);
            //可用"size( )"方法,计算向量的总个数
        for( i=0;i<MyVT.size( );i++)
            System.out.print(MyVT.elementAt(i));
        System.out.println("");
        System.out.print("Search Data:");
```

```
        //查找向量的元素
    i=MyVT.indexOf("Ja",0);
    System.out.print("The Element \"Ja\" At:"+i);

    System.out.println("");
    System.out.print("Modify Data:");
        //修改向量的元素
    MyVT.setElementAt("ja",i);
    for( i=0;i<MyVT.size();i++)
        System.out.print(MyVT.elementAt(i));
    System.out.println("");
    System.out.print("Remove Data:");
        //删除向量的元素
    MyVT.removeElementAt(MyVT.size()-1);
    for( i=0;i<MyVT.size();i++)
        System.out.print(MyVT.elementAt(i));
  }
}
```

4. 程序分析与注意事项

此部分包括程序解释及常见问题。

① "int Fibonacci[ ];" 和 "Fibonacci=new int[32];" 语句分别定义和创建了一个一维数组。此数组的数据类型为整型，其数组的大小为32。需要注意的是：创建数组时，一定要规定数组的大小，否则将会出错，其出错信息为 "**Missing array dimension**"，大意为：数组失去度，即数组没有大小。

② "Fibonacci[i] = Fibonacci[i-1]+Fibonacci[i-2];" 语句的作用是：访问数组的元素。通过改变数组的下标就可访问数组的元素。访问数组时，数组的下标从0开始到数组的个数减1。如果超过此范围，出现出错信息 "**java.lang.ArrayIndexOutOfBoundsException at ArrayVector.main(Compiled Code)**"，大意为："数组的下标越界，在ArrayVector文件的main函数中。"

③ "int[ ][ ] TriangleYH" 和 "TriangleYH=new int[10][10];" 语句分别定义和创建了一个二维数组。二维数组实质是一维数组的一维数组。即每一个一维数组的元素又为一个一维数组。访问其格式为："TriangleYH[0][0]，TriangleYH[0][1]，…，TriangleYH[1][0]，TriangleYH[1][1]，…，TriangleYH[9][8]，TriangleYH[9][9]。"

④ "Vector<String> MyVT=new Vector<String>(8,4);" 语句的作用是：创建一个向量对象。其初始大小为8，向量增量为4。

⑤ "MyVT.addElement(str[i]);" 语句的作用是：向向量中添加元素。添加元素时，系统会根据添加元素的个数判断其是否超过向量的初始大小，如果超过，系统将按向量的增量大小分配给向量空间，使得向量不至于越界。

⑥ "System. out. print(MyVT. elementAt(i)+" ");" 语句的作用是：访问向量中的元素。用 "elementAt(i)" 可以访问向量中的元素。其中参数为向量元素的位置。

⑦ "MyVT. setElementAt("J",i);" 语句的作用是：修改指定的向量的元素。其完整的格式为：向量对象名. setElementAt(String str,index i)。其中 "str" 为要替换的字符串，"i" 为向量元素的位置。

⑧ "i=MyVT. indexOf("y",0);" 语句的作用是：查找指定的字符串在向量中的位置。其完整的格式为：向量对象名. indexOf(String str,int start)。其中 "str" 为要查找的字符串，"start" 为开始查找的位置。如果没有找到，则返回-1。

⑨ "MyVT. removeElementAt(MyVT. size( )-1);" 语句的作用是：删除向量最后一个元素，其中 "MyVT. size( )" 的作用是得到向量的总个数。因向量的第一个元素的位置为0，则最后一个元素的位置为总个数-1。

练习题

1. 已知一个双精度类型的变量，将其分别转换为整型、浮点型、长整型。
2. 已知 x, y, z 三个数，比较它们的大小并按 x, y, z 从大到小的顺序排列。
3. 根据某课程的百分制成绩，要求显示对应的五等级制的评定。评定的条件为：优（大于等于90分）、良（大于等于80分小于90分）、中（大于等于70分小于80分）、及格（大于等于60分小于70分）、不及格（小于60分）。

提示：可将此成绩取10的整，这样就可用 Switch Case 语句分别列出每个等级。

4. 用 for 循环在控制台输出以下图案：

     @
    @ @
   @ @ @
  @ @ @ @

5. 用辗转相除法求两个已知的自然数 m, n 的最大公约数和最小公倍数。要求用 "do…while" 语句。

提示：

（1）对于已知的两数 m, n，使得 $m>n$；
（2）m 除以 n 得余数为 r；
（3）若 $r=0$，则 n 为得到的最大公约数，程序结束，否则执行步骤（4）；
（4）n 赋给 m，r 赋给 n，再重复执行步骤（2）；
（5）最小公倍数 $=(m*n)/$最大公约数。

6. 用循环结构分别打印九九乘法表。
7. 定义两个数组，将第一个数组中的元素复制到第二个数组中，使得第二个数组和第一个数组完全相同。再将第二个数组进行从大到小的排序，再将两个数组中的相对应元素一一进行比较。统计出相同元素的个数。
8. 创建一个向量，将其元素全部删除（可用 removeElementAll( ) 方法）。再向向量中添加元素。查找向量的指定元素,如果存在,将继续查找,直到结束,最后分别列出其所在的位置;否则显示"没有找到!"的信息。将这些找到的元素后分别插入 1,2,……元素。例如:向量的元素分别为:JS, JJJSSS, WJS, JS, Greenfir, GF, JS。此时,查找 JS 后,屏幕显示"JS"at:0,3,6。插入元素后,向量的元素为 JS, 1, JJJSSS, WJS, JS, 2, Greenfir, GF, JS, 3。

第 2 章 面向对象的程序设计（Ⅰ）
——系统类的应用

Java 是面向对象的程序设计语言，系统类是 Java 程序设计的基础，系统类贯穿了整个 Java 程序的设计。其常用的有 java.lang 类、java.io 类、java.util 类、java.net 类、java.awt 类、java.applet 类。

本章主要介绍系统中的输入输出类、数据类型类、字符串处理类、数学计算类和时间日期类的用法。

实践 2-1 输入输出类

1. 实践结果
本例的运行结果如图 2-1 所示。

```
请输入同学的姓名，以"#"结束：
张三
李四
王五
#
你的同学为：张三    李四    王五
```

图 2-1 实践 2-1 在控制台的输出

2. 实践目的
本例介绍标准的输入/输出的方式。熟悉 Java 语言在 DOS 环境下的输入/输出的方式，包括 main 函数、System 类。

（1）输入

main 函数：此方法输入是通过 main 函数的参数传递的。用 Java 编辑器编辑时在文件名后添加命令行。

read( )方法：此方法将得到用户在屏幕上输入的字符。

readLine( )方法：此方法将得到用户在屏幕上输入的字符串。

以上输入的方法的运行过程为：用户在屏幕上输入的字符或字符串，系统是以比特流方式得到的，用 BufferedReader 类的 in 对象转换为缓冲字符或字符串。

(2) 输出

print( )方法:此方法将在屏幕输出信息。此信息可以是任何一种类型的数据。

println( )方法:此方法在屏幕输出信息后,还输出一个回车换行符,使得光标移到下一行第一个字符处。

3. 程序源代码

程序的源代码如下:

```java
package exp1;

import java.io.BufferedReader;
import java.io.IOException;
import java.io.InputStreamReader;
import java.util.Vector;

public class Exp1 {

    public static void main(String[] args) {
        Vector<String> MyVT = new Vector<String>(10, 5);  //定义一个向量
        String name = "";
        System.out.print("请输入同学的姓名,以\"#\"结束:\n");
        do {
            try {
                BufferedReader in = new BufferedReader(new InputStreamReader(
                    System.in));
                name = in.readLine();
                MyVT.addElement(name);
            } catch (IOException e) {
                System.out.println("捕获异常:" + e.toString());
            }
        } while (!name.equals("#"));
        MyVT.removeElementAt(MyVT.size() - 1);
        System.out.print("你的同学为:");
        //遍历元素
        for (int i = 0; i < MyVT.size(); i++)
            System.out.print(MyVT.elementAt(i) + "    ");
        System.out.println();
    }

}
```

4. 程序分析与注意事项

此部分包括程序解释及常见问题。

以下的语句段:

```
try
{
    BufferedReader in
        =new BufferedReader(new InputStreamReader(System.in));
    name=in.readLine();
    MyVT.addElement(name);
}catch(IOException e){};
```

作用是得到用户在屏幕上输入的字符串。其中的 try…catch 是异常处理的代码块,将在以后的章节详细介绍。

实践 2-2 数据类型类及其转换

1. 实践结果

本例运行结果如图 2-2 所示。

```
1.字符类型的判断:请输入一个字符串
1qwE8g,.!
The Character(1) is Digit.
The Character(q) is LowerCase Letter.
The Character(w) is LowerCase Letter.
The Character(E) is UpperCase Letter.
The Character(8) is Digit.
The Character(g) is LowerCase Letter.
I can't recognize the character(,)!
I can't recognize the character(.)!
I can't recognize the character(!)!
共输入了9个字符!
2.数据类型的相互转换:
Float(floatValue):1234.567
Float to Integer(intValue):1234
Float to Double(doubleValue):1234.5670166015625
3.十进制转换为其他进制:
Decimal:10
Decimal to Hex(toHexString):a
Decimal to Octal(toHexString):12
Decimal to Binary(toHexString):1010
4.各数据类型的取值范围:
Use (MIN_VALUE,MAX_VALU):
-2147483648<=Integer<=2147483647
4.9E-324<=Double<=1.7976931348623157E308
1.4E-45<=Float<=3.4028235E38
-9223372036854775808<=Long<=9223372036854775807
```

图 2-2 实践 2-2 在控制台的输出

2. 实践目的

本例介绍数据类型类的应用,及其各数据类型之间的转换,包括字符型、双精度、单精度、整型、长整型类等。

(1) 字符类型类（Character）

此类提供了几种确定的一个字符的类型，并可将字符从大写转换为小写，或从小写转换为大写。还可完成大小写的判断（isLowerCase( )和 isUpperCase( )方法）、数字的判断（isDigit( )方法）、字母的判断（isLetter( )方法）。

(2) 双精度类（Double）

此类中的 floatValue( )方法、longValue( )方法、intValue( )方法分别将双精度类型的数据转换为浮点型、长整型、整型的数据。

用 toString( )方法将双精度类型的数据转换为字符串类型，而 valueOf( )方法则将字符串类型的数据转换为双精度类型的数据。

(3) 单精度类（Float）

此类的用法和双精度类的用法相似。

(4) 整型类（Integer）

此类中的 floatValue( )方法、longValue( )方法、doubleValue( )方法分别将双精度类型的数据转换为浮点型、长整型、长整型的数据。

用 toString( )方法将整型类型的数据转换为字符串类型，而用 valueOf( )方法则将字符串类型的数据转换为整型类型的数据。

parseInt( )方法：将把字符串参数解析为一个带符号的十进制数，即此方法的参数可以是一个数据类型的数据，也可是一个表达式。当为表达式，此方法将先把此表达式的结果计算出来，再转换为十进制数。

可用 toBinaryString( )方法、toOctalString( )方法、toHexString( )方法将一个十进制分别转换为二进制、八进制、十六进制数字符串。

(5) 长整型类（Long）

此类的用法和整型类的用法相似。

3. 程序源代码

程序的源代码如下：

```java
package exp2;

import java.io.BufferedReader;
import java.io.IOException;
import java.io.InputStreamReader;

public class Exp2 {

    public static void main(String[] args) throws IOException {
        char c = ' ';
        float f = 1234.567F;
        int i = 10;
        System.out.println("1. 字符类型的判断:请输入一个字符串");
        BufferedReader scanner = new BufferedReader(new InputStreamReader(
            System.in));
```

```java
String str = scanner.readLine();//从控制台读入字符串
int index = 0;                  //下标
while (index< str.length()) {
    c = str.charAt(index);
    //判断输入的字符是否为数字
    if (Character.isDigit(c))
        System.out.println("The Character(" + c + ") is Digit.");
    //判断输入的字符是否为小写字母
    else if (Character.isLetter(c) && Character.isLowerCase(c))
        System.out.println("The Character(" + c
                + ") is LowerCase Letter.");
    //判断输入的字符是否为大写字母
    else if (Character.isLetter(c) && Character.isUpperCase(c))
        System.out.println("The Character(" + c
                + ") is UpperCase Letter.");
    else {
        System.out.println("I can't recognize the character(" + c
                + ")!");
    }
    index++;
}
System.out.println("共输入了" + str.length() + "个字符!");
System.out.println("2. 数据类型的相互转换:");
System.out.print("Float(floatValue):");
System.out.println(new Float(f).floatValue());
//将浮点型转换为整型
System.out.print("Float to Integer(intValue):");
System.out.println(new Float(f).intValue());
//将浮点型转换为双精度类型
System.out.print("Float to Double(doubleValue):");
System.out.println(new Float(f).doubleValue());
System.out.println("3. 十进制转换为其他进制:");
System.out.print("Decimal:");
System.out.println(i);
//十进制转换为十六进制
System.out.print("Decimal to Hex(toHexString):");
System.out.println(Integer.toHexString(i));
//十进制转换为八进制
System.out.print("Decimal to Octal(toHexString):");
System.out.println(Integer.toOctalString(i));
//十进制转换为二进制
System.out.print("Decimal to Binary(toHexString):");
System.out.println(Integer.toBinaryString(i));
```

```java
            System.out.println("4. 各数据类型的取值范围:");
            System.out.println("Use（MIN_VALUE,MAX_VALU):");
            //整型的取值范围
            System.out.println(Integer.MIN_VALUE + "<=Integer<="
                    + Integer.MAX_VALUE);
            //双精度的取值范围
            System.out.println(Double.MIN_VALUE + "<=Double<=" + Double.MAX_VALUE);
            //浮点型的取值范围
            System.out.println(Float.MIN_VALUE + "<=Float<=" + Float.MAX_VALUE);
            //长整型的取值范围
            System.out.println(Long.MIN_VALUE + "<=Long<=" + Long.MAX_VALUE);
        }
    }
```

4. 程序分析与注意事项

此部分包括程序解释及常见问题。

① 以下语句从控制台读入一个字符串，并以此判断每个字符是否为数字、大写字母、小写字母：

```java
            BufferedReader scanner = new BufferedReader(new InputStreamReader(
                    System.in));
            String str = scanner.readLine();//从控制台读入字符串
            int index = 0;                  //下标
            while (index< str.length()) {
                c = str.charAt(index);
                index++;
            }
```

② System.out.println(new Float(f).intValue()); 语句的作用是将一个浮点类型的数据转换为整型数据。所有的数据类型转换的格式为"new 数据类型(). 数据类型 vlaue()方法"。

③ System.out.println(Integer.toHexString(i)); 语句的作用是将一个十进制数转换为十六进制数。不仅整型的数据可以转换，长整型的数据也可转换。

④ System.out.println(Integer.MIN_VALUE+"<=Integer<="+Integer.MAX_VALUE); 语句的作用是得到整型数据的取值范围。用数据类型类的 MIN_VALUE 和 MAX_VALUE 常量可取得该数据类型取值范围的最小值和最大值。

实践 2-3　字符串处理类

1. 实践结果

本例运行结果如图 2-3 所示。

2. 实践目的

本例介绍字符串处理类的应用，包括比较字符串、字符串前后缀、查找字符串、提取字符串、字符串转换。

```
1.字符串:abcdefghijklmn
2.字符串长度:14
3.字符串第0个字符为(ChartAt(0)):a
4.以第5个字符作为开头的子字符串(substring(5)):fghijklmn
5.转换为大写:ABCDEFGHIJKLMN
6.以第3个字符作为开头,以第5=(6-1)个字符作为结束的子字符串(substring(3,6)):
def
```

图 2-3　字符串处理类的应用

（1）比较字符串（compareTo( )方法和 equals( )方法）

compareTo( )方法：按字典的顺序比较两个字符串，如果相等则返回 0，如果源字符串小则返回负数，否则返回正数。格式为：

字符串 1.compareTo(字符串 2)

equals( )方法：比较两个字符串，如果相等则返回 true，否则返回 false。格式为：

字符串 1.equals(字符串 2)

（2）字符串前后缀（startsWith( )方法和 endsWith( )方法）

如果一个字符串的前缀（后缀）和指定的字符串相等，则返回 true，否则返回 false。格式为：

字符串 1.startsWith(字符串 2)
字符串 1.endsWith(字符串 2)

（3）查找字符串（indexOf( )方法）

查找指定的字符串，如果找到，返回字符串所在的位置，否则返回-1。格式为：

字符串 1.indexOf(字符串 2)

（4）提取字符串（charAt( )方法和 subString( )方法）

charAt( )方法：返回串中指定下标位置的字符，第一个字符的下标位置为 0。格式为：

字符串.charAt(字符下标位置)

subString( )方法：返回指定的下标开始位置和结束位置的字符串，第一个字符的下标位置为 0。格式为：

字符串.subString(开始位置,结束位置)

（5）字符串转换（toLowerCase( )方法和 toUpperCase( )方法）

此两种方法分别将字符串转换为大写和小写的形式。格式为：

字符串.toLowerCase( )
字符串.toUpperCase( )

3. 程序源代码

程序的源代码如下：

package exp3;

```java
public class Exp3 {
    public static void main(String[] args) {
        int i;
        String Str="abcdefghijklmn";//定义了一个字符串
        System.out.println("1.字符串:"+Str);
        //length()得到字符串的长度
        System.out.println("2.字符串长度:"+Str.length());
        //ChartAt()得到指定的字符
        System.out.print("3.字符串第0个字符为(ChartAt(0)):");
        System.out.println(Str.charAt(0));
        //substring()截取指定的子字符串
        System.out.print("4.以第5个字符作为开头的子字符串(substring(5)):");
        System.out.println(Str.substring(5));
        //toUpperCase()将字符串转换为大写字母
        System.out.print("5.转换为大写:");
        System.out.println(Str.toUpperCase());
        //substring()截取指定的子字符串
        System.out.println("6.以第3个字符作为开头,以第5=(6-1)个字符作为结束的子字符串(substring(3,6)):");
        System.out.println(Str.substring(3,6));
    }
}
```

4. 程序分析与注意事项

此部分包括程序解释及常见问题。

① "System.out.println("2.字符串长度:"+Str.length());"语句中的length()方法返回字符串（Str）的长度。

② "System.out.print("3.字符串第0个字符为(ChartAt(0)):");"语句中的chartAt()方法返回字符串（Str）中指定的下标值为0的字符。这样就可结合循环结构和length()方法提取字符串每一个字符。再根据字符类的相应方法进行处理。

③ "System.out.println(Str.substring(5));"语句的作用是返回字符串"Str"，从下标5开始到结束的字符串。

④ "System.out.println(Str.substring(3,6));"语句是substring()方法的另一种形式。其是从下标值为3开始到下标值为5结束的字符串。用此方法和循环结构结合，也可提取字符串的每个字符，但提取的字符仍为字符串形式。

⑤ "System.out.println(Str.toUpperCase());"语句的作用是将"Str"字符串中的字母全部转换为大写字母。

实践 2-4 数学类和日期时间类

1. 实践结果

本例运行结果如图 2-4 所示。

```
1. 常用的数学函数：
      E=2.718281828459045
      PI=3.141592653589793
      绝对值(y=|-2.98|)：y=2.98
      正弦(y=sin(PI/6))：y=0.49999999999999994
      开平方(y=sqrt(3))：y=1.7320508075688772
      反正弦(y=asin(Math.sqrt(3)/2))：y=1.0471975511965976
      e的x次方(y=exp(5))：y=148.4131591025766
      e的x次方(y=pow(e,5))：y=148.41315910257657
      两个数中最大者(y=max(100,200)))：y=200
2. 系统产生15个0到100随机数，15个100到200的整数：
   32.12682413357156     35.144409423062936    76.98464411891354
   9.702702901800164     64.76962460859063     69.40858226711258
   53.77817409883776     69.72518738417982     29.19595227316695
   31.122703707156962    80.07901239683761     54.579238833568326
   87.96197805123674     44.98106190657843     63.254972549857094
   163      191      157
   135      148      144
   193      189      148
   168      113      178
   199      117      183
3. 本地日期和时间：
   2018年11月7日23时21分2秒 星期3
   2018-11-07
   23:21:02.904
   2018-11-07T23:21:02.924
```

图 2-4 数学类和日期时间类的应用

2. 实践目的

掌握数学类和时间日期类的常用方法。

(1) 数学类

E 常量：返回距离 e 最近的 double 值，E=2.7182818284590452354。

PI 常量：返回距离圆周率最近的 double 值，PI=3.14159265358979323846。

abs()方法：计算参数的绝对值。

sin()方法、cos()方法、tan()方法：分别计算参数的正弦、余弦、正切。

asin()方法、acos()方法、atan()方法：分别计算参数的反正弦、反余弦、反正切。

exp()方法：返回 e 的参数次方。

floor 方法：返回不大于参数的最大 double 值。

log 方法：计算自然对数。

max()方法和 min()方法：分别返回两个数的最大值和最小值。

pow()方法：返回 a 的 b 次方。格式为：pow(a,b)。

random()方法：返回0.0到1.0之间的随机double值。
round()方法：计算距离参数最近的int值。
sqrt()方法：返回参数的开平方。
（2）日期和时间类

LocalDate类、LocalTime类、LocalDateTime类可以分别获得本地日期、本地时间、本地日期时间的对象，并利用不同的方法获取和设置年、月、日、小时、分、秒等信息。

3. 程序源代码

程序的源代码如下：

```java
package exp4;

import java.time.LocalDate;
import java.time.LocalDateTime;
import java.time.LocalTime;
import java.time.format.DateTimeFormatter;
public class Exp4 {
    public static void main(String[] args) {
        //调用数学类的方法，进行数学计算
        System.out.println("1. 常用的数学函数：");
        System.out.println("    E=" + Math.E);
        System.out.println("    PI=" + Math.PI);
        System.out.println("    绝对值(y=|-2.98|):y=" + Math.abs(-2.98));
        System.out.println("    正弦(y=sin(PI/6)):y=" + Math.sin(Math.PI / 6));
        System.out.println("    开平方(y=sqrt(3)):y=" + Math.sqrt(3));
        System.out.println("    反正弦(y=asin(Math.sqrt(3)/2)):y="
                + Math.asin(Math.sqrt(3) / 2));
        System.out.println("    e的x次方(y=exp(5)):y=" + Math.exp(5));
        System.out.println("    e的x次方(y=pow(e,5)):y=" + Math.pow(Math.E, 5));
        System.out.println("    两个数中最大者(y=max(100,200)):y=" + Math.max(100, 200));
        System.out.println("2. 系统产生15个0到100随机数，15个100到200的整数:");
        //用"random()"方法产生0到1的随机数
        //用"round()"方法提取离参数最近的整数
        for (int i = 0; i < 30; i++) {
            if (i < 15)
                System.out.print(Math.random() * 100 + "\t");
            else
                System.out.print(Math.round(Math.random() * 100 + 100) + "\t");
            if ((i + 1) % 3 == 0)
                System.out.println("");
        }
        System.out.println("3. 本地日期和时间：");
```

```java
            LocalDate ld = LocalDate.now( );         //本地日期类
            LocalTime lt = LocalTime.now( );         //本地时间类
            String str;
            //得到当前日期的年份
            str = ld.getYear( ) + "年";
            //得到当前日期的月份
            str += ld.getMonthValue( ) + "月";
            //得到当前日期
            str += ld.getDayOfMonth( ) + "日";
            //得到当前时间的小时
            str += lt.getHour( ) + "时";
            //得到当前时间的分钟
            str += lt.getMinute( ) + "分";
            //得到当前时间的秒
            str += lt.getSecond( ) + "秒";
            //得到当前日期的星期
            str += " 星期" + ld.getDayOfWeek( ).getValue( );
            System.out.println(str);                 //输入日期和时间
            //利用 DateTimeFormatter 输出日期和时间
            System.out.println( ld.format( DateTimeFormatter.ISO_DATE ) );
            System.out.println( lt.format( DateTimeFormatter.ISO_TIME ) );

    System.out.println( LocalDateTime.now( ).format( DateTimeFormatter.ISO_DATE_TIME ) );
        }
    }
```

4. 程序分析与注意事项

此部分包括程序解释及常见问题。

① "System.out.println("   E = " + Math.E);" 语句使用了数学类的域和方法。所有的数学类的域和方法都是套用一定格式就可使用。比较简单，在此不再细述。

② "System.out.print(Math.round(Math.random( ) * 100 + 100) + " \t");" 语句先提取 100 到 200 的随机数，再将得到的随机数转换为整数，以达到提取 100 到 200 的整数的目的。

③ 对于本地日期和本地时间类的各种方法，其含义简单明了，不再累述；其用法在代码中已经注释。

④ DateTimeFormatter 类用于对日期、时间信息的格式化，具体内容请参考 JavaAPI。

练习题

1. 用 main( ) 函数中命令参数的方式完成以下程序。
（1）如果命令参数的个数为 1，则提示用户下次输入用户名和 name 字符串。
（2）如果用户输入了用户名和 name 字符串，将提示此信息"【用户名】欢迎进入"。
（3）如果命令参数的个数为 2，且第二个命令参数不是 name 字符串，则提示信息"匿名登入"。

(4) 如果命令参数超过 3 个，则提示信息"不合法的输入"。

2. 用字符串输入的方式编写一个简单的通讯录。此通讯录包括姓名、性别、电话、通讯地址，并具有查询、增加、修改、删除等功能，且每次执行一次操作后将用以下的格式输出相应的信息：

编号	姓名	性别	电话	通讯地址
01	AAA	男	123456	北京
02	BBB	女	456789	福建

3. 在屏幕中输入一个包含数字、字母、符号的字符串，编写程序统计出数字、大写字母、小写字母、符号的个数。

4. 屏幕中输入一个 Java 标准的四则运算表达式。将此表达式转换为一个整型十进制数。再将此十进制数转换为二进制、八进制和十六进制。

5. 编写一个简单的文本加密程序。提示用户输入一段英语单词和一个数字密码。英语单词和数字密码用"#"号隔开，如输入"I Love you#3"，用两个变量分别存储单词和数字，再将单词逆序排列，单词将变为"uoy evoL I"。再根据数字密码的大小与字符串的长度取余运算，用得到的值将单词向右循环移动，则以上的单词将变为"L I uoy evo"。

6. 将一段英语单词的每个单词的首字母转换为大写字母。例如，转换前为"my name is small tiger brother"，转换后为"My Name Is Small Tiger Brother"。

第 3 章 面向对象的程序设计（Ⅱ）
——用户自定义类

在面向对象的程序设计中，类是一个独立的程序单位，其成员域，用来描述对象的属性；成员方法，用来描述对象的行为。类具有封装、继承和多态的特性。类的实例就是我们常说的对象。因此学习面向对象的程序设计就必须理解类的含义及其对象的特性。

本章主要介绍类的定义、对象的创建、类的继承、类的修饰、类的多态、类的封装、接口等概念及其应用。

实践 3-1 类的定义

1. 实践结果

本例运行结果如图 3-1 所示。

你好！我是汤姆
你好！我是杰瑞

图 3-1 实践 3-1 在控制台的输出

2. 实践目的

掌握类的定义和构造函数、了解面向对象的程序设计的方法。

（1）类的定义

类定义了对象的属性和行为，是对象的模板或蓝图，通常利用一个类来定义同一类型的对象。一个程序中必需有一个主类。其类名要和文件名相同。且用 Public 关键字修饰。

（2）类的成员

域：用于描述对象的属性。

方法：用于描述对象的行为。

构造函数：用于初始化对象。

（3）对象

对象是类的一个实例。当创建一个对象后，其将有类所有的属性和方法。

3. 程序源代码

程序的源代码如下：

```
package exp1;
public class Exp1 {
```

第3章 面向对象的程序设计（Ⅱ）——用户自定义类

```java
    public static void main(String[] args) {
        //TODO Auto-generated method stub
        Student tom,jerry;                          //声明对象
        tom = new Student();                        //利用空构造函数实例化tom
        tom.setName("汤姆");                        //调用属性的setter方法
        tom.setAge(2);
        tom.setId(12345);
        jerry = new Student("杰瑞",3,6789);         //调用带参数的构造函数实例化jerry
        System.out.println(tom.SayHello());         //让tom,jerry打招呼
        System.out.println(jerry.SayHello());
    }

}

class Student{

    /**
     * 姓名
     * 年龄
     * 学号
     */
    private String name;
    private int age;
    private long id;


    /**
     * 空构造函数
     */
    public Student() {
        super();
    }
    /**
     * 带参数的构造函数
     * @param name 姓名
     * @param age 年龄
     * @param id 学号
     */
    public Student(String name, int age, long id) {
        super();
        this.name = name;
        this.age = age;
        this.id = id;
```

```java
/**
 *定义一个方法用于见面打招呼
 * @return 返回一个字符串
 */
public String SayHello() {
    return "你好!"+"我是"+name;
}
public String getName() {
    return name;
}
public void setName(String name) {
    this.name = name;
}
public int getAge() {
    return age;
}
public void setAge(int age) {
    this.age = age;
}
public long getId() {
    return id;
}
public void setId(long id) {
    this.id = id;
}
}
```

4. 程序分析与注意事项

此部分包括程序解释及常见问题。

① 本程序定义两个类，一个主类 Exp1 和一个用户类 Student，主类的作用是创建对象、在浏览器中输出相应的信息，用户类是用户自定义的类。

定义类的格式为：

```
[修饰关键字] Class 类名 [继承|接口]
{
    域的声明：
    方法的声明：
    类名(参数列表){}        //构造函数的声明
}
```

定义类需要注意：一个程序只能有一个主类，且主类一定要以文件名相同，否则会出错。

② "class Student" 的作用是定义一个 Student 类。其由三个属性、一个 SayHello 方法和

两个构造函数组成。其中 public String SayHello( )是定义类成员的方法,格式为:

[修饰关键字] 数据类型 方法名(参数列表){方法体}

③ "public Student (String name, int age, long id)"用于定义类的构造函数。其作用是为此类成员初始化。其定义的注意事项:
- 构造函数的方法名与类名相同;
- 构造函数没有修饰词,即其无返回值,且不能用 void 关键字修饰;
- 如果用户没有定义构造函数,系统将自动生成一个空的构造函数"类名( )",作用是在创建对象时被调用。

④ "Student tom,jerry;"的作用是定义两个 Student 类的对象,之后分别调用带参数和不带参数构造函数,实例化这两个对象。

对象的创建的格式为:

  类名 对象名 =new 构造函数(参数列表);

或

  类名 对象名;
  对象=new 构造函数(参数列表);

对象的创建和声明数据类型的变量相似。首先用"类名 对象名"声明一个对象,再用 new 关键字为对象开辟内存空间,最后用构造函数为创建的对象初始化数据。

⑤ "return"你好!"+"我是"+name;"的作用是调用对象的属性构建一个打招呼用的字符串。当一个对象被创建时,域就成为对象的属性,方法成为对象的方法。调用对象用"."符号连接。其格式为:

  对象名.属性名 或 对象名.方法名(参数列表)

⑥ tom.SayHello( )的作用是调用对象的方法,即引用类成员的方法。

实践 3-2 类的继承

1. 实践结果

本例的运行结果如图 3-2 所示。

```
Hello  Person:人名
Hello Student: 学生名
Hello to Person: Person默认name
Hello  Person:Person默认name
```

图 3-2 实践 3-2 在控制台的输出

2. 实践目的

掌握类的继承的实现，即子类、域的继承、方法的继承、类成员的覆盖。

（1）类的继承

继承是一个类对另一个类的延续，被继承的类称为父类或超类，继承的类称为子类。

（2）域的隐藏

在子类中定义与父类同名的域，父类的域将被隐藏。所谓隐藏是指父类中的域在子类仍然存在，只不过它不能被子类直接访问，但可通过 super 关键字访问。

（3）方法的覆盖

方法的覆盖是指在子类中定义与父类完全相同的方法，父类的方法将被覆盖，但可通过 super 关键字访问。

3. 程序源代码

程序的源代码如下：

```
package exp2;
public class Exp2 {

    public static void main(String[] args) {
        //TODO Auto-generated method stub
        Person p = new Person();
        Student s = new Student();
        p.name = "人名";
        s.name = "学生名";
        p.say();
        s.say();
        s.say2();
        s.sayFromPerson();
    }

}
//Person.java
package exp2;

public class Person {
    public String name="Person 默认 name";
    public void say(){
        System.out.println("Hello  Person:"+name);
    }
}
//Student.java
package exp2;

public class Student extends Person {
```

```java
    public String name;              //与父类的 name 域同名

    public void say() {              //覆盖父类的同名方法
        System.out.println("Hello Student:" + name);
    }

    public void say2() {             //子类定义了一个新的方法
        //访问父类的 name 域
        System.out.println("Hello to Person:" + super.name);
    }

    public void sayFromPerson()      //子类定义了一个新的方法
    {   //调用父类的方法
        super.say();
    }
}
```

4. 程序分析与注意事项

此部分包括程序解释及常见问题。

① "public class Student extends Person" 语句定义了一个类 Student，其用关键字 extends 继承了 Person 类。这就是类的继承，其完整的格式为：

[修饰符] SubClassName Extends SuperClassName
{
    类体
}

② Student 类中定义了与 Person 类完全相同的 name 域。子类中定义了与父类完全相同的域，父类中的域将被隐藏，但仍可通过 super 关键字访问。例如：

"System.out.println("Hello to Person:" + super.name);"

③ Student 类中定义了与 Person 类完全相同的方法 public void say()，这样定义后，子类中的方法将具有新的行为特性。父类中的同名方法，仍可通过 super 关键字访问，例如：

"super.say();"

④ 多次出现的 super 关键字，其作用在于：super 表示当前类的直接父类，在子类中用 super 关键字就可直接访问其父类中的域和方法，这使得父类与子类的关系更为紧密。此外还有关键字 this，此关键字表示当前类本身，具体地说是当前类的另一个名字。一般来说，super 关键字在书写时可省略。

实践 3-3　类的修饰

1. 实践结果

本例运行结果如图 3-3 所示。

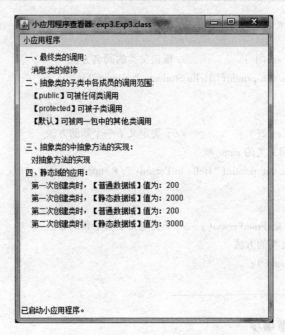

图 3-3 实践 3-3 的 Applet 运行结果

2. 实践目的

掌握类成员（域、方法）的定义方式及其访问控制符的应用。

（1）抽象（abstract）

包含抽象方法的类一定是抽象类，但抽象类可以不包含抽象方法。继承自抽象父类的非抽象子类，必须实现抽象父类的所有抽象方法。抽象类不能利用 new 关键字进行实例化，但可以定义构造函数（用于在子类中被调用）。

（2）最终（final）

利用 final 修饰类和方法后，类不能被继承，方法不能被子类重写；用 final 修饰数据域后，该数据域成为一个常数。

（3）静态（static）

static 修饰的变量被类的所有实例所共享，static 修饰的方法不依赖于类的实例。静态变量或方法可以在实例方法中使用，但静态方法中不能使用实例的变量或方法。

（4）控制符（public、private、protected）

public 关键字意为公有的，即可被任何其他的类访问。

private 关键字意为私有的，即只能被自己的类访问。

protected 关键字意为保护的，即可以被其子类访问。

如果没有使用修饰符，则仅能被同一个包中的其他类访问。

3. 程序源代码

程序的源代码如下：

```
package exp3;

public class Exp3 extends java.applet.Applet{    //定义主类，在浏览器中输出相应的信息
```

```java
        public int pubFiled;

        protected void proMethod( ) {
        }

        public void paint(java.awt.Graphics g) {
            FinalClass finalClassObject = new FinalClass("");
            g.drawString("一、最终类的调用:", 10, 20);
            g.drawString(finalClassObject.showMess( ), 20, 40);

            DefClass extendedAbstractClassObject = new DefClass( );
            g.drawString("二、抽象类的子类中各成员的调用范围:", 10, 60);
            g.drawString(extendedAbstractClassObject.pubField, 20, 80);
            g.drawString(extendedAbstractClassObject.proField, 20, 100);
            g.drawString(extendedAbstractClassObject.defField, 20, 120);

            g.drawString("三、抽象类的中抽象方法的实现:", 10, 160);
            g.drawString(extendedAbstractClassObject.AbMethod( ), 20, 180);

            g.drawString("四、静态域的应用:", 10, 200);
            ContainStaticClass containStaticObject = new ContainStaticClass( );
            g.drawString("第一次创建类时,【普通数据域】值为: " + containStaticObject.normalVar, 20, 220);
            g.drawString("第一次创建类时,【静态数据域】值为: " + ContainStaticClass.staticVar, 20, 240);
            ContainStaticClass containStaticObject2 = new ContainStaticClass( );
            g.drawString("第二次创建类时,【普通数据域】值为: " + containStaticObject2.normalVar, 20, 260);
            g.drawString("第二次创建类时,【静态数据域】值为: " + ContainStaticClass.staticVar, 20, 280);

        }
    }

    abstract class AbstClass {          //定义抽象类,作为子类的模板

        //定义抽象方法
        abstract String AbMethod( );

    }

    class DefClass extends AbstClass {   //定义默认类,此类只能在同一个包中的其他类调用
```

```java
        public String pubField = "【public】可被任何类调用";
        protected String proField = "【protected】可被子类调用";
        String defField = "【默认】可被同一包中的其他类调用";

        String AbMethod() {
            return "对抽象方法的实现";
        }
    }

    final class FinalClass {                        //定义最终类
        //定义最终域,在程序中不能赋值,相当于一个常量
        final String DEFINE = "消息:类的修饰";
        //定义私有域,此域只能在其自身类调用,其他类(包括其子类)无权调用
        private String msg;

        //定义最终方法,此方法不能被重新定义或重载
        final String showMess() {
            return msg;
        }

        FinalClass(String m) {
            msg = m == "" ? DEFINE : m;
        }
    }

    class ContainStaticClass {
        int normalVar = 100;
        //定义静态域
        static int staticVar = 1000;


        ContainStaticClass() {
            normalVar += 100;
            staticVar += 1000;
        }
    }
```

4. 程序分析与注意事项
此部分包括程序解释及常见问题。
① "public int pubFiled;" 语句定义了一个共有访问域。一个域被 public 修饰表明它可以

被任意其他类所访问和引用,如果一个类被定义为公用的,此类就可以被任意其他类引用。

② "protected String accessPriField( )" 语句定义了一个保护方法。其可被该类本身、同一个包中的其他类和其子类引用。该方法内部访问了当前类的私有数据域。

③ "private String priField" 语句定义了一个私有域,此域只能被此类自身访问和修改。所以它拥有最高的保护级,如果一个域只是被自己调用,一般定义为私有域。

④ "static int staticVar = 1000;" 语句定义了一个静态域,它是类的域,被所有当前类的对象所共享。

⑤ "abstract class AbstClass" 语句定义了一个抽象类。一个类一旦继承了一个抽象类,子类必须实现抽象类中的抽象方法。

⑥ "final class FinalClass" 语句定义了一个最终类,通常用来表示一个固定作用、用来完成特定功能的类。最终类不能被继承。

实践 3-4 类的多态

1. 实践结果

本例的运行结果如图 3-4 所示。

```
交通工具vehicle运动速度(float)=100.11!
交通工具vehicle运动速度(int)=100!
-----
交通工具vehicle运动中!
飞机plane飞往中国
-----
交通工具vehicle运动中!
汽车car飞驰中!
-----
汽车car飞驰中!
汽车car飞驰中!
```

图 3-4 实践 3-4 的控制台输出

2. 实践目的

掌握如何实现方法的重载和构造函数的重载。

(1) 方法重写

子类中使用和父类一样的方法签名和返回值,从而对方法实现细节的修改。

(2) 重载

重载是指使用同样的方法名,但是不同的方法签名类定义多个方法。既可以发生在由于继承而相关的不同类中,也可以发生在同一个类中。

(3) 多态

多态意味着可以让父类的对象指向子类的对象,在 Java 语言中对象变量是多态的。

3. 程序源代码
程序的源代码如下：

```java
//Exp4.java
package exp4;

public class Exp4 {

    public static void main(String[] args) {
        Vehicle v = new Vehicle();
        Car c = new Car();
        Plane p = new Plane();

        //同一类中的方法重载
        v.showSpeed(100.11);
        v.showSpeed(100);
        System.out.println("-----");
        //子类对父类方法重载
        v.move();
        p.move("中国");
        System.out.println("-----");
        //子类对父类方法的重写
        v.move();
        c.move();
        System.out.println("-----");
        //父类对象指向子类的对象，多态
        Vehicle v2 = new Car();
        v2.move();
        c.move();
    }
}

//Vehicle.java
package exp4;

public class Vehicle {

    public void move() {
        System.out.println("交通工具 vehicle 运动中!");
    }
    public void showSpeed(int speed) {
        System.out.println("交通工具 vehicle 运动速度(int)= "+speed+"!");
    }
    public void showSpeed(double speed) {      //showSpeed 为同一个类中的重载
```

System.out.println("交通工具 vehicle 运动速度(float) = "+speed+"!");
        }
    }

//Car.java
package exp4;

public class Car extends Vehicle {

    @Override
    public void move() {   //重写父类的方法
        System.out.println("汽车 car 飞驰中!");
    }

}

//Plane.java
package exp4;

public class Plane extends Vehicle {

    public void move(String target) {   //重载父类的方法
        System.out.println("飞机 plane 飞往"+target);
    }

}

4. 程序分析与注意事项

此部分包括程序解释及常见问题。

① Vehicle 类是 Car 和 Plane 的父类。Vehicle 中定义了一个 move 方法和两个 showSpeed 方法，其中 showSpeed 方法是在同一个类中进行的方法重载。

② 在类 Car 中重写了 Vehicle 类中的 move 方法，是子类对父类方法的重写：

    public void move() {...}

方法的重写只能发生在子类和父类之间，而且方法签名完全一致。

③ 在类 Plane 中重载了父类 Vehicle 中的 move 方法，属于父类和子类之间的方法重载。

    public void move(String target) {......}

④ 创建 Vehicle 的实例 v2，但使用 Car 类的实例对 v2 进行引用赋值，这就是多态。

    Vehicle v2 = new Car();

实践 3-5 类的封装

1. 实践结果

本例运行结果如图 3-5 所示。

圆的直径为=25.0
圆的面积为=3925.0

图 3-5 实践 3-5 的控制台输出

2. 实践目的

掌握包的使用，包括包的定义和引用。

① 类抽象的目的是将类的实现和使用分类，而隐藏类的实现细节就称为类的封装。

② 从类开发的角度看，类的封装是为了让用户能够更好地使用类，而不必关心类的内部细节。

③ 类应该通过构造函数、属性的 setter、getter 方法等手段规定类的使用方法。

3. 程序源代码

程序的源代码如下：

```java
//Exp5.java
package exp5;

public class Exp5 {
    public static void main(String[] args) {
        //TODO Auto-generated method stub
        Circle c = new Circle();

        //禁止类对象的使用者直接访问类的私有成员属性
        //System.out.println(c.radius);

        //使用者可以通过直径的 setter、getter 方法修改圆对象的直径
        //圆的半径虽然是成员属性，但被很好地隐藏起来
        c.setDiameter(50.0);
        System.out.println("圆的直径=" + c.getDiameter());

        //使用者可以得到圆的面积，但内部细节不可见
        System.out.println("圆的面积=" + c.caculateArea());
    }
}

//Circle.java
package exp5;

public class Circle {
    private double radius;                    //半径,私有成员
```

```
        public double getDiameter( ) {              //对外暴露直径的getter
            return radius;
        }

        public void setDiameter( double diameter) {  //对外暴露直径的setter
            this.radius = diameter/2;
        }

        public double caculateArea( ) {              //封装计算圆面积的方法
            return 2 * 3.14 * radius * radius;
        }
    }
```

4. 程序分析与注意事项

此部分包括程序解释及常见问题。

① 类 Circle 将半径 radius 设为私有，外部不可见。而是通过 setter、getter 方法将直径的概念暴露给外部，尽管 Circle 内部并没有直径 diameter 这一成员。

② 方法 caculateArea 封装了圆面积的计算方法，对于 Circle 类的使用者而言，这一实现细节是隐藏的，换而言之，不需要 Circle 类的使用者关心这一过程。

③ 封装体现了隐藏特点，无论是将成员隐藏，还是将方法的实现细节隐藏，最终的目的是方便类的使用者使用类，同时确保类的成员数据的安全。

实践 3-6 接口技术

1. 实践结果

本例运行结果如图 3-6 所示。

> 我是充电自行车，充电电压220V
> 我是手机，充电电压5V
> 电池在充电...
> 电池充电时间过长，爆炸了...
> 电池污染了环境...

图 3-6 实践 3-6 的控制台输出

2. 实践目的

掌握接口的相关内容，包括接口的定义、接口的继承和接口的实现。

（1）接口

接口是一种与类相似的结构，但只能包含常量和抽象方法，接口主要描述类应当具备的功能。不能通过 new 关键字实例化接口，但可以声明接口的变量，接口的变量必须引用实现了接口的类的对象。

使用 implements 关键字可使类实现接口。接口定义了一种规范,凡是实现了同一接口的类,均具有相同的行为特点。一个类能够实现多个接口,但只能有一个父类。

接口中的数据域都是 public、static、final,所以这些修饰符可以省略。接口中的方法都是 public,所以这些修饰符可以省略。Java 语言规范建议省略这些多余的关键字。

(2) 接口与抽象类

通常接口使用形容词或名词,而类常常使用名词。

抽象类和接口均用于定义一些共同的特征。抽象类常常表达了强烈的"is-a"(是一种)的含义,而接口更多表达的是"is-kind-of"(是一类)的含义。建议更多地使用接口,因为接口比较灵活。

此外,Java 不支持多重继承,而是支持实现多个接口。这样可以提供多重继承的功能,避免多重继承的复杂性。

(3) 语法规则

接口的定义:用关键字 interface 可以定义一个接口。

接口的实现:用关键字 implements 实现一个接口。

接口的继承:用关键字 extends 实现一个接口继承另一个接口。

3. 程序源代码

程序的源代码如下:

```java
//Exp6.java
package exp6;

public class Exp6 {

    public static void main(String[] args) {
        //TODO Auto-generated method stub
        //定义接口的变量,并引用接口的实现类的对象
        Rechargeable r1 = new Bicycle();
        Rechargeable r2 = new Phone();
        //接口的变量,具有一致的可充电的特性
        r1.charge();
        r2.charge();

        //定义接口的变量,并引用接口的实现类的对象
        Explosive e = new Battery();
        //Explosive 接口对 Rechargeable,Pollutional 接口进行了多继承
        //因此具备了可充电和污染性的特点,
        e.charge();
        e.explode();
        e.pollute();
    }
}
```

}
//Rechargeable.java
package exp6;

/**
 * @author Administrator 接口，表明该设备可充电
 */
public interface Rechargeable {
    int Household_Electricity = 220;
    int USB_Elecricity = 5;
    void charge();   //充电需要的方法
}
//Pollutional.java
package exp6;

/**
 * @author Administrator
 * 具有污染性的特点
 */
public interface Pollutional {
    void pollute();
}
//Explosive.java
package exp6;

/**
 * @author Administrator
 * 存在爆炸风险的特点
 * 继承自充电接口、序列化接口
 */
public interface Explosive extends Rechargeable, Pollutional {
    void explode();   //爆炸的特点
}
//Bicycle.java
package exp6;

public class Bicycle implements Rechargeable {

    @Override
    public void charge() {
        System.out.println("我是充电自行车,充电电压"+Rechargeable.Household_Electricity+"V");
```
```

```java
}

//Phone.java
package exp6;

public class Phone implements Rechargeable {
    @Override
    public void charge() {
        System.out.println("我是手机,充电电压"+Rechargeable.USB_Elecricity+"V");
    }
}
//Battery.java
package exp6;

public class Battery implements Explosive {

    @Override
    public void charge() {
        //TODO Auto-generated method stub
        System.out.println("电池在充电...");
    }

    @Override
    public void explode() {
        //TODO Auto-generated method stub
        System.out.println("电池充电时间过长,爆炸了...");
    }

    @Override
    public void pollute() {
        //TODO Auto-generated method stub
        System.out.println("电池污染了环境...");
    }

}
```

4. 程序分析与注意事项
此部分包括程序解释及常见问题。

① "interface Rechargeable" 语句定义了一个接口。其完整的格式为:

[修饰符] interface 接口名称 [extends 父接口1,父类接口2…]
{

        //接口体
    }

定义接口和定义类很相似，但接口可继承多个父接口。

② "public interface Explosive extends Rechargeable，Pollutional" 语句的作用是使接口 Explosive 多重继承了 Rechargeable 和 Pollutional 接口。因此 Explosive 接口的成员为：

    public void charge( )
    public void explode( )
    public void pollute( )

③ "public class Phone implements Rechargeable" 语句实现了 Rechargeable 接口。其完整的格式为：

    class 类名 implements 接口名 1[，接口名 2…]
    {
        //类体
    }

当实现一个接口时，必须实现接口中的方法。

练习题

1. 定义一个处理盒子的类。其成员包括盒子的长、宽、高域，计算盒子的体积、表面积、设置并得到盒子的长、宽、高的方法。再用构造函数初始化盒子的长、宽、高的值。
2. 在第 1 题中，如果要处理 10 个盒子的相关属性，如何创建对象？

提示：可用对象数组创建。格式为：

    类名 对象名[ ]=new 类名[对象的个数]
    对象名[0]=new 构造函数(参数列表)
    …
    对象名[对象的个数-1]=new 构造函数(参数列表)

以上三行的内容可用一个循环实现。

3. 思考如何使一个类成员方法有多个返回值？

提示：先定义一个类（成为类型类）作为类成员方法的类型，然后返回"类型类"的对象即可。

4. 在子类中，如何用父类的构造函数初始化父类的域，又如何用子类的构造函数初始化父类的域？

5. 定义方法，完成以下功能。

第一步：使得调用此方法时，其实际参数可以是整型，或是浮点型，或是字符型，或是字符串型，或是布尔型，或是双精度类型。

第二步：定义多个这样的方法，应如何定义？

提示：可用类的多态实现。如果是多个这样的方法，可以把此方法作为一个固定的类。

定义其他方法时,方法的参数类型为固定的类的类型即可。即对象作为方法的参数的应用。以下是本题第二步的参考,请读者认真体会。

```java
public class PracticeAT extends java.applet.Applet
{
    public void paint(java.awt.Graphics g)
    {
        DealClass myDC = new DealClass();
        myDC.method1(new TypeClass(10));
        myDC.method1(new TypeClass("I Love you"));
        myDC.method1(new TypeClass(100.001));
        /*以上方法的方法名相同,而实参的数据类型不同。这是第一步要做的内容。以下又
          定义了"method2"和"method3"方法。它们也是方法名相同,而实参的数据类型不
          同。*/
        myDC.method2(new TypeClass('c'));
        myDC.method2(new TypeClass(100));
        myDC.method3(new TypeClass(true));
        myDC.method3(new TypeClass("Tiger's wife"));
    }
}

class DealClass
{
    int vi;float vf;char vc;
    String vs,SS;boolean vb;double vd;
    void method1(TypeClass TX){}
    void method2(TypeClass TX){}
    void method3(TypeClass TX){}
}

class TypeClass
{
    int vi;float vf;char vc;String vs;boolean vb;double vd;
    nt getInt(){return vi;}
    float getFloat(){return vf;}
    char getChar(){return vc;}
    String getString(){return vs;}
    boolean getBoolean(){return vb;}
    double getDouble(){return vd;}
    TypeClass(int t){vi=t;}
    TypeClass(float t){vf=t;}
    TypeClass(char t){vc=t;}
    TypeClass(String t){vs=t;}
```

```
    TypeClass(boolean t){vb=t;}
    TypeClass(double t){vd=t;}
}
```

6. 一个类中定义了一个私有域，另一个类如何访问此私有域？

提示：只能在定义私有域的类中定义一个方法，返回此私有域的值。

7. 用"package js1.js2.js3"定义一个包，包中有"jsclass1"和"jsclass2"公有类。以下能访问"jsclass1"类的语句有：

（1）import js1.*;

（2）import js1.js2.js3;

（3）import js1.js2.js3.*;

（4）import js1.js2.js3.jsclass2;

（5）import js1.js2.js3.jsclass1;

8. 以下程序是否有错误？为什么？如果有错误，如何更改。

提示："!"号的为注意点。

```
interface jsif
{
    int i;                          //!
    int j=100;                      //!
    int getvalue(){};               //!
    void setvalue(int vi);
}

interface jsep extends jsif
{
    i=100;                          //!
    int j=100;                      //!
    int getvalue()                  //!
    {
        return i;
    }
}

class jsip1 implements jsif
{
    int getvalue()                  //!
    {
        return i;
    }
                                    //!
}
interface jsip2 implements jsif     //!
```

```
    }
        int j;
        int getvalue( )
        {
            return i;
        }
        viod setvalue( int vi )         //!
        {
            i=vi;                        //!
            j=vi;                        //!
        }
    }
```

第 4 章 图形用户界面设计（Ⅰ）

通过图形用户界面，用户和程序之间可以方便友好地进行交互。

本章主要介绍 JFrame，基本几何图形的绘制，图像的现实，事件处理基础，常用界面组件和键盘事件，鼠标事件。

实践 4-1 JFrame 基础

1. 实践结果

本例运行结果如图 4-1 所示。

图 4-1 一个简单的 JFrame 应用

2. 实践目的

掌握 JFrame 类的基本用法和自定义组件的使用方法。

（1）JFrame 类的基本用法

class BlankFrame extends JFrame：创建一个自定义的 JFrame 子类，生成程序的框架。

setDefaultCloseOperation 方法：设置关闭框架式的默认动作。

setTitle 方法：设置框架的标题。

setLocationByPlatform 方法：由窗口系统定位框架。

setVisible 方法：显示框架。

add 方法：向框架添加一个组件。

pack 方法：利用组件的首选大小，调整窗口的大小。

（2）自定义组件的使用方法

class MessageComponent extends JComponent：创建一个自定义的组件类。

protected void paintComponent（Graphics g）方法：覆盖父类中的方法，用于绘制组件。

Graphics g 可以被转换成 Graphics2D 对象 g2，并利用 g2 绘制字符串。

public Dimension getPreferredSize() 方法：返回组件的首选大小，是一个 Dimension 对象。

3. 程序源代码

程序的源代码如下：

```java
package exp1;

import java.awt.*;

import javax.swing.*;

public class HelloJFrame {
    public static void main(String args[]) {
        EventQueue.invokeLater(() -> {
            BlankFrame frame = new BlankFrame(400, 500);   //实例化一个框架
                frame.setDefaultCloseOperation(JFrame.EXIT_ON_CLOSE);//定义关闭时的动作
                frame.setTitle("Hello JFrame!");           //设置框架标题
                frame.setLocationByPlatform(true);         //由窗口系统定位框架
                frame.setVisible(true);                    //显示框架
        });
    }
}


class BlankFrame extends JFrame {
    //继承自 JFrame
    public BlankFrame(int width, int height) throws HeadlessException {
        super();
        this.setSize(width, height);                       //框架的尺寸
        this.add(new MessageComponent());                  //向框架添加一个组件
        this.pack();                                       //利用组件的首选大小，调整窗口的大小
    }
}

/**
 * @author Administrator 自定义的显示消息的组件
 */
class MessageComponent extends JComponent {
    public static final int DEFAULT_W = 300;                //组件的首选尺寸
    public static final int DEFAULT_H = 200;

    @Override
    protected void paintComponent(Graphics g) {             //绘制组件的方法
```

```
        //TODO Auto-generated method stub
        Graphics2D g2 = (Graphics2D) g;                    //转换 Graphics2D
        g2.drawString("I am a message!", 100, 100);        //在(100,100)位置绘制一个字符串
    }

    @Override
    public Dimension getPreferredSize() {                  //组件首选尺寸大小
        //TODO Auto-generated method stub
        return new Dimension(DEFAULT_W, DEFAULT_H);
    }

}
```

4. 程序分析与注意事项

此部分包括程序解释及常见问题。

① paintComponent 方法用于重绘组件，但一定不要自己调用该方法，这个方法由程序自动调用。

② 若想强制刷新屏幕，应当调用 repaint 方法，而不是 paintComponent 方法。

③ 所有 Swing 组件必须由时间分配程序进行配置，线程负责把鼠标和键盘控制转给组件，所以 main 方法中才有如下代码：

```
EventQueue.invokeLater(() -> {
    ...
});
```

④ g2.drawString 方法中的位置信息（100，100）代表了字符串的位置，字符串开始位置位于水平 100 像素点，字符串基线位置位于垂直 100 像素点。

实践 4-2　基本几何图形的绘制

1. 实践结果

本例运行结果如图 4-2 所示。

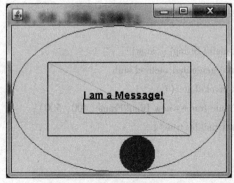

图 4-2　绘制基本的几何图形

2. 实践目的

了解绘图的基本方法和图形用户界面的绘图原理。

(1) 绘图基本步骤

创建图形：创建 Java2D 库中的 Rectangle2D、Ellipse2D、Line2D 等对象。

设置颜色：调用 Graphics2D 的 setPaint 方法设置当前颜色。

绘图或填充：调用 Graphics2D 的 draw 方法绘制图形，或者利用 fill 方法进行填充。

(2) 字体使用

如果希望使用特定的字体绘制字符，则需要利用字体名、风格、大小创建一个 Font 对象。字体名可以是系统中已经安装的字体的名称，风格主要是指常规、加粗、斜体或加粗斜体（Font 类中常量对应为 PLAIN，BOLD，ITALIC，BOLD+ITALIC），大小为一个整数，代表字体的点数目大小（72 个点/英寸）。

3. 程序源代码

程序的源代码如下：

```
package exp2;

import java.awt.Color;
import java.awt.Dimension;
import java.awt.EventQueue;
import java.awt.Font;
import java.awt.Graphics;
import java.awt.Graphics2D;
import java.awt.HeadlessException;
import java.awt.font.FontRenderContext;
import java.awt.geom.Ellipse2D;
import java.awt.geom.Line2D;
import java.awt.geom.Rectangle2D;

import javax.swing.JComponent;
import javax.swing.JFrame;

public class Exp2 {

    public static void main(String[] args) {
        //TODO Auto-generated method stub
        EventQueue.invokeLater(()->{
            BlankFrame frame=new BlankFrame(400, 500);
            frame.setVisible(true);
        });
    }
}
```

```java
class BlankFrame extends JFrame {
    //继承自 JFrame
    public BlankFrame(int width, int height) throws HeadlessException {
        super();
        this.setSize(width, height);              //框架的尺寸
        this.add(new MessageComponent());         //向 JFram 添加一个组件
        this.pack();                              //利用组件的首选大小,调整窗口的大小
    }
}

/**
 * @author Administrator 自定义的显示消息的组件
 */
class MessageComponent extends JComponent {
    public static final int DEFAULT_W = 300;      //组件的首选尺寸
    public static final int DEFAULT_H = 200;

    @Override
    protected void paintComponent(Graphics g) {   //绘制组件的方法
        //TODO Auto-generated method stub
        String str="I am a Message!";
        Graphics2D g2 = (Graphics2D) g;           //转换 Graphics2D
        //设置系统中已安装的字体
        Font font=new Font("SansSerif",Font.BOLD,14);
        g2.setFont(font);
        g2.drawString(str, 100, 100);             //在(100,100)位置绘制一个字符串
        //对字符串进行测量,获得其外部矩形尺寸,并绘制
        FontRenderContext frc=g2.getFontRenderContext();
        Rectangle2D fontRect=font.getStringBounds(str, frc);
        //矩形的上边界为字符串的基线位置
        Rectangle2D surrrect = new Rectangle2D.Double(100.0, 100.0, fontRect.getWidth(), fontRect.getHeight());
        g2.draw(surrrect);
        //设置颜色,利用椭圆的外界矩形绘制椭圆,
        g2.setPaint(Color.RED);
        Ellipse2D ellipse=new Ellipse2D.Double(0,0,300,200);//创建椭圆
        g2.draw(ellipse);
        //重新设置颜色,绘制一个矩形
        Rectangle2D rect=new Rectangle2D.Double(50,50,200,100);
        g2.setPaint(Color.BLACK);
        g2.draw(rect);
        //重新设置颜色,绘制一条线
```

```
            Line2D line=new Line2D.Double(50,50,250,150);
            g2.setPaint(Color.GREEN);
            g2.draw(line);
            //利用 RGB 方法生成新的颜色,并填充一个圆
            Ellipse2D circle=new Ellipse2D.Double(150,150,50,50);//创建圆
            g2.setPaint(new Color(0,138,128));
            g2.fill(circle);
        }

        @Override
        public Dimension getPreferredSize() {        //组件首选尺寸大小
            //TODO Auto-generated method stub
            return new Dimension(DEFAULT_W, DEFAULT_H);
        }
    }
```

4. 程序分析与注意事项

此部分包括程序解释及常见问题。

① 绘制图形时,Graphics 对象所使用的单位为像素,组件表面左上角点为 (0,0),坐标轴 X 的数值水平向右增大,Y 的数值垂直向下增大。

② drawString 方法中的字符串起始位置的坐标代表了字符串基线的左侧端点位置。

③ Rectangle2D.Double 等类似的构造函数表明需要使用 Double 的数据作为参数。

④ Color 类用于定义颜色,除了 Color 类中预定义的 13 个常量用于表示颜色外,还可以使用 Color 的构造函数自定义颜色,例如 new Color(int red,int green,int blue)。此外,在 SystemColor 类中也定义了很多颜色的名字。

⑤ 对于字体的使用,除了使用系统中已经安装的字体外,还可以利用 TrueType 格式的字体文件,方法是读取字体文件为输入流,然后利用 Font.createFont 静态方法创建一个字体。

实践 4-3 图像的显示

1. 实践结果

本例介绍图像显示的方式和图像设置的相关属性。其运行结果如图 4-3 所示。

2. 实践目的

在 Swing 中显示图像是比较容易实现的,读取图像文件,创建 Image 对象后,调用 Graphics2D 的 drawImage 方法绘制即可。

3. 程序源代码

程序的源代码如下:

第4章 图形用户界面设计（Ⅰ）

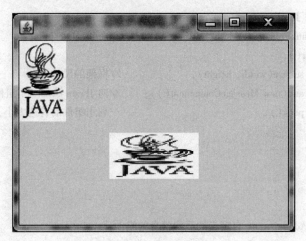

图 4-3 显示图像

```
package exp3;

import java.awt.Dimension;
import java.awt.EventQueue;
import java.awt.Graphics;
import java.awt.Graphics2D;
import java.awt.HeadlessException;
import java.awt.Image;
import javax.swing.ImageIcon;
import javax.swing.JComponent;
import javax.swing.JFrame;




public class Exp3 {

    public static void main(String[] args) {
        //TODO Auto-generated method stub


        //TODO Auto-generated method stub
        EventQueue.invokeLater(()->{
            BlankFrame frame=new BlankFrame(400, 500);
            frame.setVisible(true);
        });


    }

}class BlankFrame extends JFrame {
```

```java
            //继承自 JFrame
            public BlankFrame(int width, int height) throws HeadlessException {
                super();
                this.setSize(width, height);              //框架的尺寸
                this.add(new MessageComponent());         //向 JFram 添加一个组件
                this.pack();                              //利用组件的首选大小,调整窗口的大小
            }
        }

        /**
         * @author Administrator 自定义的显示消息的组件
         */
        class MessageComponent extends JComponent {
            public static final int DEFAULT_W = 300;      //组件的首选尺寸
            public static final int DEFAULT_H = 200;

            @Override
            protected void paintComponent(Graphics g) {   //绘制组件的方法
                //TODO Auto-generated method stub

                Graphics2D g2 = (Graphics2D) g;           //转换 Graphics2D
                Image img=new ImageIcon("javalogo.gif").getImage();
                g2.drawImage(img,0,0,null);               //图像不进行缩放
                g2.drawImage(img, 100, 100,100,50, null); //将图像绘制到指定区域,会进行缩放
            }

            @Override
            public Dimension getPreferredSize() {         //组件首选尺寸大小
                //TODO Auto-generated method stub
                return new Dimension(DEFAULT_W, DEFAULT_H);
            }
        }
```

4. 程序分析与注意事项

此部分包括程序解释及常见问题。

g2.drawImage(img,0,0,null)语句中绘制的图像不进行缩放。

g2.drawImage(img, 100, 100,100,50, null)语句将图像绘制到指定区域,会进行缩放。

drawImage 方法的最后一个参数为 ImageObserver 类型,可以为 null。

实践 4-4　事件处理基础

1. 实践结果
本例运行结果如图 4-4 所示。

图 4-4　实践 4-4 的运行界面

2. 实践目的
了解事件处理的基本概念和流程。

（1）AWT 事件处理机制

事件源：用于产生事件，例如按钮产生 ActionEvent 对象，窗口产生 WindowEvent 对象。
监听器：监听器是实现了特定接口的类的实例，可以注册到事件源，从而监听事件。
事件的传递：事件发生时，事件源将事件对象传递给已经注册的监听器。
事件的处理：监听器对象中有特定的方法处理事件。

（2）适配器类

为了便于简化监听器接口的实现，很多 AWT 监听器接口都有一个对应的适配器类，又称为 Adapter。适配器类虽然实现了接口中的方法，但什么也没处理。需要说明的是，监听器接口中若仅存在一个方法，则没有对应的 Adapter 类，此时常常利用 lambda 表达式进行监听器方法的实现。

（3）常用的简单对话框

Swing 中 JOptionPane 有四个用于显示对话框的方法，可以方便地显示对话框，这四个静态方法分别是：showMessageDialog、showConfirmDialog、showOptionDialog 和 showInputDialog。

3. 程序源代码
程序的源代码如下：

```
package exp4;
import java.awt.EventQueue;
```

```java
import java.awt.HeadlessException;
import java.awt.event.WindowAdapter;
import java.awt.event.WindowEvent;

import javax.swing.JButton;
import javax.swing.JFrame;
import javax.swing.JOptionPane;
import javax.swing.JPanel;
public class Exp4 {
    public static void main(String[] args) {
        //TODO Auto-generated method stub
        EventQueue.invokeLater(()->{
            ButtonFrame frame=new ButtonFrame(400,500);
            frame.setVisible(true);
        });
    }
}
class ButtonFrame extends JFrame {
    //继承自 JFrame
    private JPanel panel=null;              //JPanel 用于放置按钮
    public ButtonFrame(int width, int height) throws HeadlessException {
        super();
        for(int i=0;i<5;i++) {              //添加 5 个按钮
            this.addButton("按钮"+i);
        }
        this.setSize(width, height);        //框架的尺寸
        this.add(panel);                    //向 JFram 添加一个组件
        this.pack();                        //利用组件的首选大小,调整窗口的大小
        this.addWindowListener(new WindowAdapter() {   //使用匿名类
            //WindowAdapter 是 WindowListener 的一个实现类,但没有做任何事情
            @Override
            public void windowClosing(WindowEvent e) {
                //处理窗体关闭的事件
                //TODO Auto-generated method stub
                super.windowClosing(e);
                //显示一个消息对话框
                JOptionPane.showMessageDialog(panel,"用户关闭窗口。。。。");
            }
        });
    }
    private void addButton(String text) {
        if(this.panel==null)                //如果没有初始化 panel,则初始化
            this.panel=new JPanel();
```

```
            JButton button = new JButton(text);    //添加普通的文字按钮
            this.panel.add(button);
            button.addActionListener((event)->{
                //添加按钮的监听器，使用 lambda 表达式
                //显示一个消息对话框
                JOptionPane.showMessageDialog(panel, "你单击了:按钮2 按钮"+text);
            });
        }
    }
```

4. 程序分析与注意事项

此部分包括程序解释及常见问题。

this.addWindowListener(new WindowAdapter(){...});语句向 BlankFrame 的实例添加了窗口事件 WindowEvent 的监听器。这里使用了匿名类，且该匿名类扩展自 WindowAdapter 类，覆盖了其中的 WindowClosing 方法。

函数式接口：只有一个抽象方法的接口被称为函数式接口。当需要这种接口对象时，可以提供一个 lambda 表达式。

button.addActionListener((event)->{...});语句向 button 对象添加了 ActionEvent 事件的监听器，这个监听器需要实现接口 ActionListener。而 ActionListener 接口仅含一个方法 actionPerformed，因此这里利用了 lambda 表达式。

实践 4-5 　常用界面组件和键盘事件

1. 实践结果

本例运行结果如图 4-5 所示。

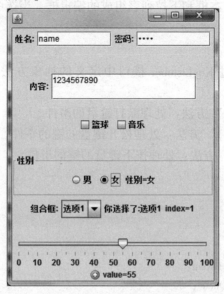

图 4-5　Swing 中常见的组件

2. 实践目的

了解单文本域、密码域、文本区、标签、单选框、复选框、组合框、滑块的基本用途，以及它们之间的区别，了解其选择事件的用法。

（1）文本域（JTextField）和密码域（JPasswordField）

用户可以在组件内输入文本，密码域为了防止密码泄露而采用回显字符遮盖用户输入。常用方法为 setText 和 getText，可以设置和获取文本。

（2）标签（JLabel）

用于文本指示附近组件的功能和作用，可以使用 SwingConstants 接口中的常量来配置标签的排列方式，也可以利用 setIcon 方法为标签设置图标。

（3）文本区（JTextArea）

常用于接收输入超过一行的文本，用户利用 Enter 键换行，每一行结尾为'\n'。利用 setRows 和 setColumns 方法可以修改其容纳的行数和列数。必要时利用 setLineWrap 方法可以开启换行特性，此时超过显示范围的文本自动换行，但这种自动换行不会增加'\n'的个数，文本的内容没有变化。很多时候可以和滚动窗格 JScrollPane 配合使用，由 JScrollPane 来自动生成滚动条。JTextArea 只能编辑无格式文本，如果编辑格式化文本，需要使用 JEditorPane 类。

（4）复选框（JCheckbox）

复选框是表示选中状态的组件，外观是一个矩形和文字描述，当选中时矩形内出现一个对钩。

（5）单选按钮（JRadioButton）

单选按钮通常成组出现，利用 ButtonGroup 对单选按钮进行编组，组内的单选按钮互斥。

（6）组合框（JComboBox）

组合框是一种有一个文本框和下拉列表组成的编辑框，当选中其中的一项时，文本显示在文本框中。

（7）键盘事件的响应

键盘事件的响应是通过 KeyListener 接口中定义的抽象方法处理的，该接口包括以下方法。

"keyPressed(KeyEvent)"方法：处理敲打键盘的事件。

"keyReleased(KeyEvent)"方法：处理松开键盘的键的事件。

"keyTyped(KeyEvent)"方法：处理按下键盘的键的事件。

3. 程序源代码

程序的源代码如下：

```
package exp5;

import java.awt.BorderLayout;
import java.awt.EventQueue;
import java.awt.GridLayout;
```

```java
import java.awt.HeadlessException;
import java.awt.Image;
import java.awt.event.KeyAdapter;
import java.awt.event.KeyEvent;
import java.awt.event.WindowAdapter;
import java.awt.event.WindowEvent;

import javax.swing.BorderFactory;
import javax.swing.ButtonGroup;
import javax.swing.Icon;
import javax.swing.ImageIcon;
import javax.swing.JCheckBox;
import javax.swing.JComboBox;
import javax.swing.JComponent;
import javax.swing.JFrame;
import javax.swing.JLabel;
import javax.swing.JOptionPane;
import javax.swing.JPanel;
import javax.swing.JPasswordField;
import javax.swing.JRadioButton;
import javax.swing.JScrollPane;
import javax.swing.JSlider;
import javax.swing.JTextArea;
import javax.swing.JTextField;
import javax.swing.SwingConstants;

/**
 * @author Administrator
 *
 */
public class Exp5 {
    public static void main(String[] args) {
        //TODO Auto-generated method stub
        EventQueue.invokeLater(() -> {
            ButtonFrame frame = new ButtonFrame(400, 500);
            frame.setDefaultCloseOperation(JFrame.EXIT_ON_CLOSE);
            frame.setVisible(true);
        });
    }
}

class ButtonFrame extends JFrame {
    //继承自 JFrame
```

```java
        private JPanel panel = null;                    //JPanel 用于放置按钮

    public ButtonFrame(int width, int height) throws HeadlessException {
        super();
        this.setSize(width, height);                    //框架的尺寸
        this.setLayout(new BorderLayout());
        panel = new JPanel();
        panel.setLayout(new GridLayout(0, 1));          //最外层 panel 为网格布局,单列

        //创建一个文字域面板
        JPanel fieldPanel = new JPanel();               //JPanel 默认流式布局
        fieldPanel.add(new JLabel("姓名:", JLabel.RIGHT));
        JTextField nameField = new JTextField("name", 10);
        nameField.addKeyListener(new KeyAdapter() {     //为文本域增加监听器

            @Override
            public void keyPressed(KeyEvent event) {
                //TODO Auto-generated method stub
                super.keyPressed(event);
                System.out.println("KeyChar=" + event.getKeyChar() + " KeyCode=" + event.getKeyCode() + " KeyText=" + event.getKeyText(event.getKeyCode()));
            }

        });
        fieldPanel.add(nameField);
        fieldPanel.add(new JLabel("密码:", JLabel.LEFT));
        fieldPanel.add(new JPasswordField("pass", 10));
        panel.add(fieldPanel);
        //创建一个文本区面板
        JPanel areaPanel = new JPanel();
        areaPanel.add(new JLabel("内容:", JLabel.CENTER));
        JScrollPane scrollPane = new JScrollPane(new JTextArea("1234567890", 2,
                20));                                   //放入滚动面板,自动产生滚动条
        areaPanel.add(scrollPane);
        panel.add(areaPanel);

        //创建一个复选框面板
        JPanel checkPanel = new JPanel();
        checkPanel.add(new JCheckBox("篮球"));
        checkPanel.add(new JCheckBox("音乐"));
        panel.add(checkPanel);

        //创建一个单选框面板
```

```java
JPanel radioPanel = new JPanel();
JLabel radioValue = new JLabel("性别=");
ButtonGroup group = new ButtonGroup();              //使用按钮组完成互斥
JRadioButton rb1 = new JRadioButton("男");
rb1.addActionListener((event) -> {
    radioValue.setText("性别=" + rb1.getText());
});
rb1.setSelected(true);
JRadioButton rb2 = new JRadioButton("女");
rb2.addActionListener((event) -> {
    radioValue.setText("性别=" + rb2.getText());
});
group.add(rb1);                                     //添加到按钮组
group.add(rb2);
radioPanel.add(rb1);                                //添加到面板
radioPanel.add(rb2);
radioPanel.setBorder(BorderFactory.createTitledBorder("性别"));  //使用题目边框
radioPanel.add(radioValue);
panel.add(radioPanel);

//创建一个组合框面板
JLabel comboValue = new JLabel("你选择了:");
JPanel comboPanel = new JPanel();
comboPanel.add(new JLabel("组合框:"));
JComboBox<String> combo = new JComboBox<String>();
for (int i = 0; i < 10; i++) {
    combo.addItem("选项" + i);
}
combo.addActionListener((event) -> {
    comboValue.setText("你选择了:" + combo.getSelectedItem() + "  index="
            + combo.getSelectedIndex());
});
comboPanel.add(combo);
comboPanel.add(comboValue);
panel.add(comboPanel);

//创建一个滑块面板
JLabel sliderValue = new JLabel("Value", SwingConstants.CENTER);
ImageIcon img = new ImageIcon("add-icon.gif");           //设置 label 的 Icon 图标
sliderValue.setIcon(img);
JPanel sliderPanel = new JPanel();
sliderPanel.setLayout(new BorderLayout());
JSlider hslider = new JSlider(0, 100, 30);
```

```
hslider.addChangeListener((event) -> {
    sliderValue.setText("value=" + hslider.getValue());
});
hslider.setPaintLabels(true);              //显示标签
hslider.setMajorTickSpacing(10);           //大标记间隔
hslider.setMinorTickSpacing(5);            //小标记间隔
hslider.setPaintTicks(true);               //显示标记
sliderPanel.add(hslider, BorderLayout.NORTH);      //北侧
sliderPanel.add(sliderValue, BorderLayout.SOUTH);  //南侧
panel.add(sliderPanel);

this.add(panel, BorderLayout.NORTH);       //向 JFram 添加一个组件
this.pack();        //利用组件的首选大小,调整窗口的大小
    }

}
```

4. 程序分析与注意事项
此部分包括程序解释及常见问题。

（1）布局

在最外层 JPanel 采用了网格布局 panel.setLayout(new GridLayout(0, 1));，单列,行数待定。后续采用了很多 JPanel 包含不同的组件,这些 JPanel 采用默认的流式布局。滑块面板 sliderPanel 使用了边框布局 sliderPanel.setLayout(new BorderLayout())。如果需要更加复杂的布局要求,可以使用网格组布局即 GridBagLayout。

（2）事件处理

对姓名文本域 nameField 采用 KeyAdapter 进行按键事件 KeyEvent 的监听,将字符打印到控制台。KeyAdapter 是一个实现了 KeyListener 接口的类,我们可以简单地继承它,然后覆盖所需的方法。对于两个 JRadioButton 和 JComboBox 组件的 ActionEvent 事件进行了监听,利用 lambda 表达式进行了事件的处理。对 JSlider 组件的 ChangeEvent 事件进行监听,利用 lambda 表达式进行了事件的处理。

（3）单选按钮组

属于同一组的单选按钮需要添加到同一个 ButtonGroup 类的对象,才能够实现互斥。代码如下：

　　group.add(rb1);group.add(rb2);

（4）标签组件

JLabel 的组件利用 setIcon 方法,可以为标签设置图标。

（5）滑块组件

JSlider 组件可以使用标签和标记进行装饰,setPaintLabels(true)用于显示标签,setPaintTicks(true)用于显示标记。

（6）KeyEvent 的常用方法

getKeyText()方法：返回按键的文本内容。如回车键值为"Enter",小键盘的数字键 1

的值为"NumPad-1"等。

getKeyCode( )方法：返回按键的键码，可和 getKeyText( )一起使用。例如，Shift 键的键码为 16。

getKeyChar( )方法：返回按键的 Unicode 字符，如果一个键没有 Unicode 字符键，则返回空。例如，键盘的"1"值为"1"，而小键盘上的"1"值也为"1"。

实践 4-6　鼠标事件

1. 实践结果

本例运行结果如图 4-6 所示。

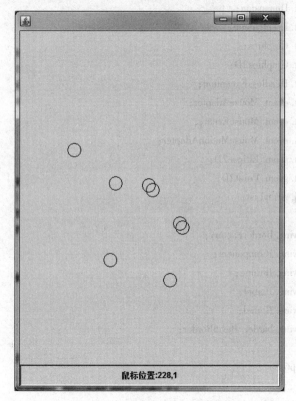

图 4-6　实践 4-6 的运行界面

2. 实践目的

熟悉鼠标事件的处理方法、相应事件接口。

（1）鼠标操作事件的处理

MouseListener 接口中定义了以下可以处理鼠标操作的抽象方法。

mouseClicked(MouseEvent)方法：处理单击鼠标的事件。

mouseEntered(MouseEvent)方法：处理鼠标进入容器的事件。

mouseExited(MouseEvent)方法：处理鼠标退出容器的事件。

mousePressed(MouseEvent)方法：处理按下鼠标键的事件。

mouseReleased(MouseEvent)方法：处理松开鼠标键的事件。
(2) 鼠标运动事件的处理
MouseMotionListener 接口中定义了以下可以处理鼠标运动的抽象方法。
mouseMoved(MouseEvent)方法：处理鼠标移动的事件。
mouseDragged(MouseEvent)方法：处理鼠标拖动的事件。

3. 程序源代码

程序的源代码如下：

```java
package exp6;

import java.awt.BorderLayout;
import java.awt.Dimension;
import java.awt.EventQueue;
import java.awt.Graphics;
import java.awt.Graphics2D;
import java.awt.HeadlessException;
import java.awt.event.MouseAdapter;
import java.awt.event.MouseEvent;
import java.awt.event.MouseMotionAdapter;
import java.awt.geom.Ellipse2D;
import java.awt.geom.Point2D;
import java.util.ArrayList;

import javax.swing.BorderFactory;
import javax.swing.JComponent;
import javax.swing.JFrame;
import javax.swing.JLabel;
import javax.swing.JPanel;
import javax.swing.border.BevelBorder;

public class Exp6 {

    public static void main(String[] args) {
        //TODO Auto-generated method stub
        EventQueue.invokeLater( ( ) -> {
            ButtonFrame frame = new ButtonFrame(400, 500);
            frame.setDefaultCloseOperation(JFrame.EXIT_ON_CLOSE);
            frame.setVisible(true);
        });
    }
}

class ButtonFrame extends JFrame {
```

```java
//继承自 JFrame
private JPanel infoPanel = new JPanel();

private SimpleDrawComponent drawComponent = new SimpleDrawComponent(400,500);
private JLabel positionLabel=new JLabel("我显示鼠标位置信息..",JLabel.LEFT);

public ButtonFrame(int width, int height) throws HeadlessException {
    super();
    this.setSize(width, height);            //框架的尺寸
    this.setLayout(new BorderLayout());     //框架采用边框布局

    //向组件添加另一个鼠标运动监听器,用于显示鼠标位置
    drawComponent.addMouseMotionListener(new MouseMotionCapture());

    infoPanel.setSize(400, 100);
    infoPanel.setBorder(BorderFactory
        .createBevelBorder(BevelBorder.LOWERED));   //斜面边框

    infoPanel.add(positionLabel);
    this.add(drawComponent, BorderLayout.CENTER);
    this.add(infoPanel, BorderLayout.SOUTH); //向 JFrame 添加一个组件
    this.pack();                             //利用组件的首选大小,调整窗口的大小
}

//内部类
class MouseMotionCapture extends MouseMotionAdapter {   //仅支持运动和拖放

    @Override
    public void mouseMoved(MouseEvent arg0) {
        //TODO Auto-generated method stub
        super.mouseMoved(arg0);
        positionLabel.setText("鼠标位置:"+arg0.getX()+","+arg0.getY());
    }

}
}
class SimpleDrawComponent extends JComponent{        //自定义组件
    private int w,h;                                 //首选大小
    private ArrayList<Ellipse2D> circles;            //保存圆
    private Ellipse2D currentCircle;                 //选中圆
```

```java
public SimpleDrawComponent(int w,int h){
    //初始化各种内部对象
    super();
    this.w=w;
    this.h=h;
    circles=new ArrayList<Ellipse2D>();
    currentCircle=null;
    MouseCapture capture=new MouseCapture();
    this.addMouseListener(capture);          //指定鼠标操作的监听
    this.addMouseMotionListener(capture);    //输定鼠标运动的监听
}
@Override
protected void paintComponent(Graphics g){   //绘制组件方法
    //TODO Auto-generated method stub
    super.paintComponent(g);
    Graphics2D g2=(Graphics2D)g;
    for(Ellipse2D c:circles){                //绘制圆
        g2.draw(c);
    }
}
@Override
public Dimension getPreferredSize(){         //返回组件首选大小
    //TODO Auto-generated method stub
    //返回首选大小
    return new Dimension(w,h);
}

//内部类,为了访问外部类的变量和方法
class MouseCapture extends MouseAdapter{     //支持所有的鼠标事件

    @Override
    public void mouseClicked(MouseEvent arg0){   //处理单击事件
        //TODO Auto-generated method stub
        super.mouseClicked(arg0);

        Point2D p=arg0.getPoint();
        System.out.println("鼠标单击"+p);
        Ellipse2D c=new Ellipse2D.Double(p.getX()-10,p.getY()-10,20,20);
        circles.add(c);
        repaint();                           //重绘组件
    }
```

```java
            @Override
            public void mouseMoved(MouseEvent e) {            //移动中找最近的圆
                //TODO Auto-generated method stub
                double dis = 20;                              //距离初始化为圆的半径
                double temp = 0;
                Point2D p = e.getPoint();
                for(Ellipse2D c:circles) {                    //找距离最近的圆
                    if(c.contains(p)) {
                        temp=Point2D.distance(c.getCenterX(), c.getCenterY(), p.getX(), p.getY());
                        if(temp<dis) {
                            dis = temp;                       //找到一个更近的圆;
                            currentCircle = c;                //保存这个
                        }
                    }
                }
            }
            @Override
            public void mouseDragged(MouseEvent e) {          //处理拖放
                //TODO Auto-generated method stub
                if(currentCircle!=null) {                     //改变圆的位置
                    currentCircle.setFrame(e.getPoint().getX(), e.getPoint().getY(), 20, 20);
                    repaint();                                //重绘组件
                }
                System.out.println("鼠标拖放..."+e.getPoint());
            }
        }
    }
```

4. 程序分析与注意事项

此部分包括程序解释及常见问题。

内部类 MouseCapture 和 MouseMotionCapture：在类 SimpleDrawComponent 内部定义，目的是使用它的内部对象。MouseMotionCapture 类定义为内部类，也是基于此目的。

JComponent 的 addMouseListener 方法和 addMouseMotionListener 方法，用于向组件分别添加鼠标操作的监听器和鼠标运动的监听器。

MouseAdapter 类是 MouseListener 接口的实现类，覆盖了所有方法，但其中的方法实现却没有做任何事情，因此编程时可以继承 MouseAdapter 类，并覆盖所需要的方法。

练习题

1. 编写程序，对图像进行分割和垂直翻转。提示：要将图像进行垂直翻转，只要将图像水平地分割成若干块，再将这些块按水平方向上逆序放置即可。

2. 编写一个简单的调色板。

3. 编辑程序，在文本框中输入一个英语单词，文本域中就添加一个英语单词。左边的组合框中是26个大写的英文字母。选择左边组合框的一项，右边组合框将显示所有以左边组合框中所选的字母开头的单词。

4. 编辑一个用户绘制的按钮。完成一般按钮所具有的功能，即鼠标按下时，按钮处于凹陷状；当鼠标放开时，按钮处于突起的状态，且按钮上的文字随着鼠标状态的改变而改变。

第 5 章 图形用户界面设计（Ⅱ）

本章主要介绍常用布局、动作和菜单、选项对话框和自定义对话框、文件选择器和颜色选择器。

实践 5-1 常用布局

1. 实践结果

本例运行结果如图 5-1、图 5-2、图 5-3 所示。

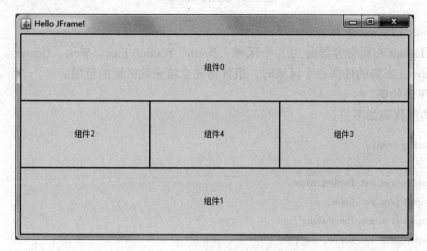

图 5-1 BorderLayout 布局

图 5-2 FlowLayout 布局

2. 实践目的

学会使用 GridLayout、FlowLayout 和 BorderLayout 布局绘图。

（1）GridLayout 布局

GridLayout 布局采用一种类似表格的行列排列方式，但每个单元大小都是一样的。new GridLayout(0,2) 构造函数中需要制定行数和列数，行数和列数的值可以为 0，但不能同时为 0。组件加入时，按照先行后列的顺序排列组件。

（2）FlowLayout 布局

当添加组件时，会按照从左向右的顺序排列。空间不够时，自动换到下一行。

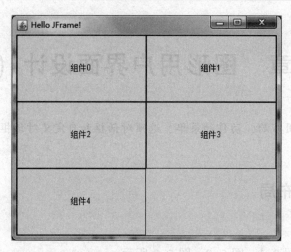

图 5-3 GridLayout 布局

(3) BorderLayout 布局

BorderLayout 布局将容器分为 5 个区域：North、South、East、West、Center。当组件加入 BorderLayout 布局的任何一个区域时，组件将完全填充此区域的范围。

3. 程序源代码

程序的源代码如下：

```
package exp1;

import java.awt.BorderLayout;
import java.awt.Color;
import java.awt.Dimension;
import java.awt.EventQueue;
import java.awt.FlowLayout;
import java.awt.Font;
import java.awt.Graphics;
import java.awt.Graphics2D;
import java.awt.GridLayout;
import java.awt.HeadlessException;
import java.awt.geom.Rectangle2D;

import javax.swing.BorderFactory;
import javax.swing.JComponent;
import javax.swing.JFrame;
import javax.swing.JPanel;

public class Exp1 {

    public static void main(String[] args) {
        //TODO Auto-generated method stub
```

```java
        EventQueue.invokeLater(() -> {
            BlankFrame frame = new BlankFrame(600, 800);    //实例化一个框架
              frame.setDefaultCloseOperation(JFrame.EXIT_ON_CLOSE);    //定义关闭时的动作
              frame.setTitle("Hello JFrame!");              //设置框架标题
              frame.setLocationByPlatform(true);            //由窗口系统定位框架
              frame.pack();
              frame.setVisible(true);                       //显示框架
        });

    }

}

class BlankFrame extends JFrame {
    int w, h;

    SimpleComponent[] components = new SimpleComponent[8];

    public BlankFrame(int w, int h) throws HeadlessException {
        super();

        this.w = w;
        this.h = h;
        this.setLayout(new GridLayout(0,2));            //网格布局
        //this.setLayout(new FlowLayout());             //流式布局
        //this.setLayout(new BorderLayout());           //边框布局
        for (int i = 0; i < 5; i++) {
            components[i] = new SimpleComponent(200, 100, "组件" + i);
            //绘制组件边界
            components[i].setBorder(BorderFactory.createLineBorder(Color.BLUE));
        }

        if (this.getLayout() instanceof BorderLayout) {     //边框布局时
            this.add(components[0], BorderLayout.NORTH);
            this.add(components[1], BorderLayout.SOUTH);
            this.add(components[2], BorderLayout.WEST);
            this.add(components[3], BorderLayout.EAST);
            this.add(components[4], BorderLayout.CENTER);
        } else {                                            //其他情况
            for (int i = 0; i < 5; i++) {
                this.add(components[i]);
            }
        }
```

```java
class SimpleComponent extends JComponent {
    int w, h;
    String name;

    public SimpleComponent(int w, int h, String name) {
        super();
        this.w = w;
        this.h = h;
        this.name = name;
    }

    /**
     * 在组件的中心绘制字符串
     * @param g2 Graphics2D 对象
     * @param str 内容
     */
    private void drawStringAtCenter(Graphics2D g2, String str) {
        Font f = g2.getFont();                                       //获得字体
        //测量字符串外围矩形
        //高度=上坡度+下坡度+行间距
        //宽度=字符串水平宽度
        Rectangle2D r = f.getStringBounds(str, g2.getFontRenderContext());
        int width = this.getSize().width;                            //组件大小
        int height = this.getSize().height;
        int x = (int)(width - r.getWidth()) / 2;                     //计算绘制起点的 X 坐标
        int y = (int)((height - r.getHeight()) / 2 - r.getY());
        //起点的 Y 坐标,由于从基线计算,所以需要追加上坡度的值
        //上坡度的值存储在字符串外围矩形的左上角点 Y 坐标(但它是负的)
        g2.drawString(str, x, y);      //绘制字符串是的坐标是基线的左侧顶点
    }

    @Override
    protected void paintComponent(Graphics g) {
        //TODO Auto-generated method stub
        super.paintComponent(g);
        Graphics2D g2 = (Graphics2D) g;
        this.drawStringAtCenter(g2, name);   //调用自定义方法,在组件的正中央绘制文字
    }
}
```

```
@Override
public Dimension getPreferredSize() {
    //TODO Auto-generated method stub
    return new Dimension(w, h);
}
```

4. 程序分析与注意事项

此部分包括程序解释及常见问题。

① JFrame 类对象的 setLayout 方法用于设置布局类型,add 方法用于添加组件,当使用 BorderLayout 时需要使用 add 的另一种重载形式,例如 .add(components[2], BorderLayout.WEST)。

② 自定义了方法 drawStringAtCenter 利用 FontRenderContext 对象和 Font 对象测量字符串的外围矩形,从而计算 drawString 所需要的起点坐标。需要注意的是,字符串绘制是从其基线的左侧端点开始的,所以需要准确计算基线位置。字符串外围矩形的高度包含上坡度、下坡度、行间距;宽度为字符串水平宽度。上坡度为字符串基线到坡顶的距离,下坡度为字符串基线到坡底的距离,行间距为某一行的坡底到下一行坡顶的距离。

③ setBorder 方法用于给组件添加特定的边框。

实践 5-2　动作和菜单

1. 实践结果

本例运行结果如图 5-4 所示。

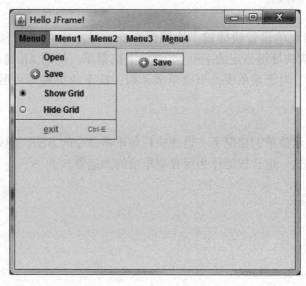

图 5-4　菜单和动作

2. 实践目的

了解 Action 动作、下拉菜单和弹出式菜单的创建和应用。

(1) Action（动作）

在用户界面中，用户可以单击按钮、按下键盘、选择菜单项来执行同一个特定的功能。为了简化编程过程，Swing 中提供了一个简洁的机制，即 Action（动作）接口。利用这一接口，可以将所有事件连接到同一个监听器，从而实现统一的处理。Action 接口封装了下列对象：命令的描述（文本和图标）和执行命令需要的参数。Action 接口有一个实现类 AbstractAction，它实现了除 actionPerformed 方法之外的所有方法。通常直接继承自 AbstractAction 以实现自定义的 Action 类。

(2) 菜单条（JMenuBar）

菜单条是下拉菜单的载体，位于当前窗口的标题栏的下边。

(3) 下拉菜单（JMenu）

下拉菜单是菜单项的载体，下拉菜单有名称。

(4) 弹出式菜单（JPopupMenu）

弹出式菜单和下拉菜单相似，只是创建后弹出式菜单不可见，只有通过一定的操作才能将其显示，在 Windows 和 Linux 系统中这一操作为单击鼠标右键。弹出式菜单需要组件调用 setComponentPopupMenu 方法设置才能在特定组件中生效。

(5) 菜单项（JMenuItem）

菜单项位于下拉式菜单或弹出式菜单中，用户单击后可以执行一个具体的命令。除了普通的菜单项，还有复选框菜单项（JCheckBoxMenuItem）和单选框菜单项（JRadioButtonMenuItem），可以在菜单项文本的旁边显示单选和复选按钮。

(6) 分隔线

分隔线的作用是将不同的菜单项进行分组，使菜单更为美观、清晰。addSeparator( ) 方法用于向菜单添加分隔线。

(7) 快捷键

可以为菜单项和菜单设置快捷键，快捷键会自动显示在菜单项和菜单的标题文字中（前提是标题文字包含快捷键指定的字母）。对于下拉菜单，使用 Alt 键+快捷键可以选择菜单条中的某一个菜单。对于菜单项，当菜单显示后，按下快捷键即可触发菜单命令（与鼠标单击效果一致）。

(8) 加速器

加速器是在不显示菜单的情况下，迅速执行菜单项命令的方法，例如我们熟悉的 Ctrl+S 键一样，很多程序将这一组合按键作为保存菜单项的加速器。

3. 程序源代码

程序的源代码如下：

```
package exp2;

import java.awt.BorderLayout;
import java.awt.Color;
import java.awt.Dimension;
```

```java
import java.awt.EventQueue;
import java.awt.GridLayout;
import java.awt.HeadlessException;
import java.awt.event.ActionEvent;

import javax.swing.AbstractAction;
import javax.swing.Action;
import javax.swing.ActionMap;
import javax.swing.BorderFactory;
import javax.swing.ButtonGroup;
import javax.swing.Icon;
import javax.swing.ImageIcon;
import javax.swing.InputMap;
import javax.swing.JButton;
import javax.swing.JCheckBoxMenuItem;
import javax.swing.JComponent;
import javax.swing.JFrame;
import javax.swing.JMenu;
import javax.swing.JMenuBar;
import javax.swing.JMenuItem;
import javax.swing.JOptionPane;
import javax.swing.JPanel;
import javax.swing.JPopupMenu;
import javax.swing.JRadioButtonMenuItem;
import javax.swing.KeyStroke;
import javax.swing.event.MenuEvent;
import javax.swing.event.MenuListener;

public class Exp2 {

    public static void main(String[] args) {
        //TODO Auto-generated method stub
        EventQueue.invokeLater(() -> {
            BlankFrame frame = new BlankFrame(600, 800);  //实例化一个框架
                frame.setDefaultCloseOperation(JFrame.EXIT_ON_CLOSE); //定义关闭时的动作
                frame.setTitle("Hello JFrame!");        //设置框架标题
                frame.setLocationByPlatform(true);       //由窗口系统定位框架
                frame.pack();
                frame.setVisible(true);              //显示框架
        });
    }
}
```

```java
class BlankFrame extends JFrame {
    int w, h;

    private JMenuBar menuBar;
    private JMenu[] menus = new JMenu[5];
    private boolean isSaved=false;
    public BlankFrame(int w, int h) throws HeadlessException {

        super();

        this.w = w;
        this.h = h;
        this.setLayout(new BorderLayout());              //边框布局
        JPanel panel = new JPanel();
        panel.setPreferredSize(new Dimension(400, 400)); //手动改变 JPanel 首选大小
        this.add(panel, BorderLayout.CENTER);
        char[] mnemonics = {'a','b','c','d','e'};        //快捷键字符
        menuBar = new JMenuBar();
        for (int i = 0; i < 5; i++) {                    //创建顶级菜单
            menus[i] = new JMenu("Menu" + i);
            //菜单名称中如果包含快捷键字符,会自动显示
            menus[i].setMnemonic(mnemonics[i]);          //设置快捷键,配个 Alt 使用
            menuBar.add(menus[i]);                       //向菜单条加入菜单
        }



        //编辑 menus0 菜单
        JMenuItem openItem = new JMenuItem("Open");
        openItem.addActionListener((event) -> {
            JOptionPane.showMessageDialog(this, "单击了 Open 菜单项!");
        });
        menus[0].add(openItem);                          //菜单尾部插入菜单项

        //利用匿名类创建 saveAction,这是一个 Action 接口的变量
        //Action 又称为动作,封装了命令的说明和执行命令所需要的参数
        Action saveAction = new AbstractAction("Save", new ImageIcon(
                "add-icon.gif")) {
            @Override
            public void actionPerformed(ActionEvent e) {
                //TODO Auto-generated method stub
                JOptionPane.showMessageDialog(panel, "执行了"
```

```java
                    + getValue(Action.NAME) + "动作");
            isSaved = true;
        }
    };

    JMenuItem saveItem = menus[0].add(saveAction);

    panel.add(new JButton(saveAction));                //利用 Action 创建按钮

    //创建弹出式菜单
    JPopupMenu popMenu = new JPopupMenu();
    popMenu.add(saveAction);
    popMenu.add(new JMenuItem("复制.."));
    panel.setComponentPopupMenu(popMenu);              //设置弹出式菜单

    //将动作对象 saveAction 添加到击键中
    InputMap inputMap = panel
        .getInputMap(JComponent.WHEN_ANCESTOR_OF_FOCUSED_COMPONENT);  //获得组件的输入
                                                                      //映射
    inputMap.put(KeyStroke.getKeyStroke("ctrl S"), "panel.save");     //在输入映射中关联
                                                                      //击键和名称
    ActionMap actionMap = panel.getActionMap();        //获得组件的动作映射
    actionMap.put("panel.save", saveAction);           //在动作映射中关联名称和动作关联

    menus[0].addSeparator();                           //加入分隔符

    //创建单选按钮菜单项
    JRadioButtonMenuItem showGridItem = new JRadioButtonMenuItem(
            "Show Grid");
    JRadioButtonMenuItem hideGridItem = new JRadioButtonMenuItem(
            "Hide Grid");
    showGridItem.setSelected(true);
    ButtonGroup bg = new ButtonGroup();
    bg.add(showGridItem);
    bg.add(hideGridItem);
    menus[0].add(showGridItem);
    menus[0].add(hideGridItem);

    //创建复选框按钮菜单
    JCheckBoxMenuItem saveFlagItem = new JCheckBoxMenuItem("已保存");
    saveFlagItem.addActionListener((event) -> {
        isSaved = !isSaved;
```

```java
                    //单击后取反
});
menus[0].add(saveFlagItem);


menus[0].addSeparator();                           //加入分隔符

JMenuItem exitItem=new JMenuItem("exit",'E');      //菜单项快捷键,菜单显示后使用
exitItem.addActionListener((event)->{
    System.exit(0);                                //退出程序
});
exitItem.setAccelerator(KeyStroke.getKeyStroke("ctrl E"));//设置加速键,不需要打开
                                                          //菜单
menus[0].add(exitItem);

//设置 menu0 菜单的监听器
menus[0].addMenuListener(new MenuListener(){       //创建一个匿名类

    @Override
    public void menuSelected(MenuEvent e){
        //TODO Auto-generated method stub
        System.out.println("菜单 0,selected!");
        if(isSaved){//根据是否执行了 save 操作,判断是否禁用退出菜单项
            exitItem.setEnabled(true);
        }
        else{
            exitItem.setEnabled(false);
        }
    }

    @Override
    public void menuDeselected(MenuEvent e){
        //TODO Auto-generated method stub
        System.out.println("菜单 0,deselected!");
    }

    @Override
    public void menuCanceled(MenuEvent e){
        //TODO Auto-generated method stub
        System.out.println("菜单 0,canceled!");
    }
});
setJMenuBar(menuBar);                              //设置 Frame 的菜单条
```

        }
    }

4. 程序分析与注意事项

此部分包括程序解释及常见问题。

程序包含一个菜单条 menuBar，利用循环创建了 5 个下拉菜单，并通过 add 方法添加到 menuBar。程序包含一个 JPanel 组件 panel，并利用 setComponentPopupMenu 方法为其设置了一个弹出式菜单，此外还向其添加了一个按钮。调用 JFrame 的 setMenuBar 方法向窗体添加菜单条。

（1）下拉菜单

5 个下拉菜单利用 setMnemonic 方法均设置了快捷键，快捷键存储在 mnemonics 数组中。

（2）自定义 Action

创建了 Action 接口的变量 saveAction，用于对保存这一动作进行统一的监听处理。saveAction 实际上是一个继承自 AbstractAction 的匿名类的实例，并设置了动作的名称和图标，这一匿名类的 actionPerformed 方法也被覆盖，具体功能为显示一个消息框，并设置 isSaved 为 true。

（3）利用 saveAction 创建组件

saveAction 初始化后，可以直接传递给 JButton 类的构造函数和菜单的 add 方法，从而创建一个与 saveAction 相关联的按钮和菜单项。

　　　　panel. add(new JButton(saveAction))
　　　　popMenu. add(saveAction)
　　　　JMenuItem saveItem = menus[0]. add(saveAction) ;

（4）单选按钮菜单项

单选按钮菜单和单选按钮的基本用法相同，同组得到单选按钮需要加入到同一 ButtonGroup 实例进行管理。

（5）复选框菜单项

复选框菜单项和复选框适用法基本相同。

（6）菜单项的快捷键和加速器

在利用 JMenuItem 的构造函数创建菜单项时，可以传入第二个参数用于指定快捷键，例如 JMenuItem exitItem = new JMenuItem("exit", 'E')。调用菜单项的 setAccelerator(KeyStroke. getKeyStroke("ctrl E")) 方法可以指定加速器。

（7）MenuListener 接口

该接口中定义了三个方法：menuSelected、menuDeselected、menuCanceled 对应不同菜单事件的处理。调用菜单的 addMenuListener 方法可以向菜单添加监听器。

实践 5-3 选项对话框和自定义对话框

1. 实践结果

本例运行结果如图 5-5 和图 5-6 所示。

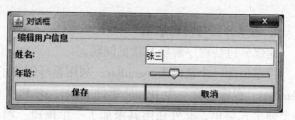

图 5-5 利用下拉列表框激活选项对话框　　图 5-6 自定义对话框

2. 实践目的

了解选项对话框的用法,以及创建自定义对话框的常用方法。

(1) JOptionPane 对话框

JOptionPane 类中有一套简单的选项对话框,利用如下静态方法可以显示这些对话框。

showMessageDialog:显示消息,用户只能单击 OK 按钮。无返回值。

showConfirmDialog:显示消息,用户可以单击 OK 或 Cancel 按钮。返回值为唯一整数,代表用户选择的结果。

```
if (sel1 == JOptionPane.YES_OPTION) {…}
```

showOptionDialog:显示消息,用户可以定制选项,例如 Object[ ] options = new Object[ ]{"String",new ImageIcon("add-icon.gif")}。用户可以单击不同选项。返回值为唯一整数,代表用户选择的结果。

```
if (sel1 == JOptionPane.YES_OPTION) {…}
```

showInputDialog:显示消息,用户可以输入一行文本。返回值为输入的文本。

(2) 对话框的有模式与无模式

当用户设置对话框的模式为"有模式"时,运行对话框就阻止其他窗口的交互。只有关闭对话框时,其他窗口才可被使用;当用户设置的对话框为"无模式"时,运行对话框用户可继续操作其他的窗口。

(3) 自定义对话框

本例展示了用两种方法进行对话框的定制,第一种是让类继承自 JDialog;第二种是定义一个普通的组件类,然后放到 JDialog 的一个普通实例中展示即可。

继承 JDialog 类:MineDialog extends JDialog,类 MineDialog 继承自 JDialog,因此其本身就是一个对话框,本例加入了一个 JComboBox 控件,并根据用户输入调用 JOptionPane 的不同静态方法分别显示消息对话框、确认对话框、选项对话框和输入对话框。调用 setVisible(true) 使对话框可见。

继承 JPanel 类:class CustomerEditor extends JPanel,CustomerEditor 类是一个 JPanel 的子类,可以放置任何控件到这个 panel 中。当需要显示为对话框时,利用一个 JDialog 对象作为这个 panel 的拥有者即可。调用 JDialog 对象的 setVisible(true) 使对话框可见。

3. 程序源代码

程序的源代码如下:

```java
package exp3;

import java.awt.BorderLayout;
import java.awt.Dimension;
import java.awt.EventQueue;
import java.awt.GridLayout;
import java.awt.HeadlessException;

import javax.swing.BorderFactory;
import javax.swing.ImageIcon;
import javax.swing.JButton;
import javax.swing.JComboBox;
import javax.swing.JDialog;
import javax.swing.JFrame;
import javax.swing.JLabel;
import javax.swing.JMenu;
import javax.swing.JMenuBar;
import javax.swing.JMenuItem;
import javax.swing.JOptionPane;
import javax.swing.JPanel;
import javax.swing.JSlider;
import javax.swing.JTextField;

public class Exp3 {

    public static void main(String[] args) {
        //TODO Auto-generated method stub
        EventQueue.invokeLater(() -> {
            BlankFrame frame = new BlankFrame(600, 800);      //实例化一个框架
            frame.setDefaultCloseOperation(JFrame.EXIT_ON_CLOSE);//定义关闭时的动作
            frame.setTitle("Hello JFrame!");                  //设置框架标题
            frame.setLocationByPlatform(true);                //由窗口系统定位框架
            frame.pack();
            frame.setVisible(true);                           //显示框架
        });
    }
}

class BlankFrame extends JFrame {
    int w, h;

    private JMenuBar menuBar;
```

```java
public BlankFrame(int w, int h) throws HeadlessException {

    super();

    this.w = w;
    this.h = h;
    this.setLayout(new BorderLayout());           //边框布局
    JPanel panel = new JPanel();                   //增加一个 Panel
    panel.setPreferredSize(new Dimension(400, 400));//手动改变 JPanel 首选大小
    this.add(panel, BorderLayout.CENTER);          //panel 放中间

    menuBar = new JMenuBar();                      //菜单条

    //第一个菜单, 用于显示 JOptionPane 中的对话框
    JMenu optionPaneMenu = new JMenu("选项对话框");
    optionPaneMenu.setMnemonic('o');//设置快捷键
    //增加一个菜单项
    JMenuItem openDialogItem = optionPaneMenu.add("显示一个选项对话框...");
    openDialogItem.addActionListener((event) -> {  //加入监听器
        JDialog d = new MineDialog(this);          //使用一个自定义对话框, MineDialog
                                                    //是 JDialog 子类
        d.setLocationRelativeTo(this);             //使用相对于当前组件的定位
        d.setVisible(true);                        //使对话框可见
    });
    menuBar.add(optionPaneMenu);

    //创建一个客户实例
    Customer customer = new Customer("张三", 20);

    //创建第二个菜单, 用于展示复杂的数据交互对话框
    JMenu dataExchangeMenu = new JMenu("数据交互的复杂的对话框");

    //增加一个菜单项, 并设置监听器
    JMenuItem customerEditItem = dataExchangeMenu.add(new JMenuItem(
        "编辑客户信息"));
    customerEditItem.addActionListener((event) -> {
        CustomerEditor ce = new CustomerEditor();   //创建客户编辑器, 这是一个自定义
                                                    //的 JPanel 子类
        ce.loadCustomer(customer);                  //加载数据
        if (ce.showMe(this)) {                      //显示对话框, 并根据用户操作进行保存
            Customer c = ce.saveCustomer();
            customer.setAge(c.getAge());
```

```java
                customer.setName(c.getName());
            }
            System.out.println(customer);
        });
        menuBar.add(dataExchangeMenu);
        setJMenuBar(menuBar);
    }
}

class MineDialog extends JDialog {                                  //直接继承 JDialog 创建自定义对话框
    public MineDialog(JFrame owner) {
        //父类构造方法,传入拥有者、标题、是否模态
        super(owner, "自定义对话框", true);

        //创建一个 JPanel,并放置一个 JCombox 组件
        JPanel panel = new JPanel();
        JComboBox<String> combo = new JComboBox<String>();
        combo.addItem("MessageDialog");
        combo.addItem("ConfirmDialog");
        combo.addItem("OptionDialog");
        combo.addItem("InputDialog");
        //根据用户选择的选项,展示不同的 JOptionPane 对话框
        combo.addActionListener((event) -> {
            int index = combo.getSelectedIndex();
            switch (index) {
            case 0:
                JOptionPane.showMessageDialog(this, combo.getSelectedItem(),
                        "消息对话框", JOptionPane.WARNING_MESSAGE);
                break;
            case 1:
                int sel = JOptionPane.showConfirmDialog(this,
                        combo.getSelectedItem(), "确认对话框",
                        JOptionPane.YES_NO_OPTION);
                if (sel == JOptionPane.YES_OPTION) {                //判断用户选择
                    System.out.println("单击了 YES!");
                } else {
                    System.out.println("单击了 NO!");
                }
                break;
```

```java
                case 2:
                    Object[] options = new Object[]{"String",
                            new ImageIcon("add-icon.gif")};       //自定义 Options
                    int sel1 = JOptionPane.showOptionDialog(this,   //父组件
                            combo.getSelectedItem(),                //消息
                            "选项对话框",                            //题目
                            JOptionPane.YES_NO_OPTION,              //选项类型
                            JOptionPane.INFORMATION_MESSAGE,        //消息类型
                            null, //亦可以单独指定图标 new ImageIcon("add-icon.gif")替换
                                  //INFORMATION_MESSAGE,
                            options,                                //选项数组
                            options[0]                              //默认选项
                            );
                    if (sel1 == JOptionPane.YES_OPTION) {           //判断用户选择
                        System.out.println("单击了字符串,index=" + sel1);
                    } else {
                        System.out.println("单击了图标,index=" + sel1);
                    }
                    break;
                case 3:
                    String s = JOptionPane.showInputDialog(this,
                            combo.getSelectedItem(), "输入对话框");//获取用户输入
                    System.out.println("用户输入了:" + s);
                    break;
                }
            });
            //设置 Panel 首选大小,并加入当前的 JDialog
            panel.add(combo);
            panel.setPreferredSize(new Dimension(200, 200));
            this.add(panel, BorderLayout.CENTER);
            this.setSize(300, 200);                                 //设置对话框大小
        }
    }

    class Customer {                                                //客户类
        private String name;
        private int age;

        public String getName() {
            return name;
        }
```

```java
    public void setName(String name) {
        this.name = name;
    }

    public int getAge() {
        return age;
    }

    public void setAge(int age) {
        this.age = age;
    }

    public Customer(String name, int age) {
        super();
        this.name = name;
        this.age = age;
    }

    @Override
    public String toString() {
        //TODO Auto-generated method stub
        return "姓名=" + name + " 年龄=" + age;
    }

}

/**
 * @author Administrator
 * 这是一个自定义的对话框类,但没有继承自 JDialog,
 * 而是继承自 JPanel
 * 其中放置一个 JTextField 编辑客户姓名
 * 一个 JSlider 编辑客户年龄
 * 两个 JButton,允许用户保存和取消
 */
class CustomerEditor extends JPanel {
    private JTextField name;
    private JSlider age;
    private JButton save, cancel;
    //定义一个 JDialog 对象,作用是在显示 CustomerEditor 时对其进行包装
    private JDialog dialog;
    //用户操作的标志
    private boolean userSaved = false;
```

```java
public CustomerEditor() {
    super();
    //初始化各种内部控件,添加响应的事件监听器
    setBorder(BorderFactory.createTitledBorder("编辑用户信息"));
    name = new JTextField();
    age = new JSlider(0, 100, 1);
    setLayout(new GridLayout(3, 2));                    //网格布局
    //创建各种组件并添加
    add(new JLabel("姓名:"));
    add(name);
    add(new JLabel("年龄:"));
    add(age);
    save = new JButton("保存");
    cancel = new JButton("取消");
    //按钮的监听器
    save.addActionListener((event) -> {
        userSaved = true;
        dialog.setVisible(false);
    });
    cancel.addActionListener((event) -> {
        userSaved = false;
        dialog.setVisible(false);
    });
    add(save);
    add(cancel);
}

public void loadCustomer(Customer customer) {
    age.setValue(customer.getAge());                    //加载数据
    name.setText(customer.getName());
}

public Customer saveCustomer() {
    return new Customer(name.getText(), age.getValue());//获取数据并封装
}

public boolean showMe(JFrame owner) {
    //显示这个自定义对话框,使用JDialog包装一下
    //将panel添加到JDialog
    userSaved = false;
    dialog = new JDialog(owner, true);
    dialog.add(this);
```

```
            dialog.pack();                                    //利用首选大小调整尺寸
            dialog.setTitle("对话框");
            //设置对话框的默认按钮
            dialog.getRootPane().setDefaultButton(cancel);
            dialog.setLocationRelativeTo(owner);              //使用相对于拥有者的定位方式
            dialog.setVisible(true);                          //显示对话框
            return this.userSaved;                            //返回用户操作标志
        }
    }
```

4. 程序分析与注意事项

此部分包括程序解释及常见问题。

setLocationRelativeTo 方法可以利用与其他组件的相对位置，来定位当前的窗格或对话框。

JOptionPane.showOptionDialog 的方法可以使用定制选项的方式，选项参数应当是一个 Object 对象数组，例如 Object[] options = new Object[]{"String", new ImageIcon("add-icon.gif")}，当数组中的对象是 String、Icon、Component 类型的实例时，将创建字符串按钮、图标按钮和显示组件，当使用其他对象时，会使用对象的 toString 方法作为按钮的文字；返回值则是被选中的选项的索引值，如果用户没有选择而是关闭，返回值则为 CLOSED_OPTION。

当需要设置自定义对话框的默认按钮时，可使用 JDialog 对象的 getRootPane().setDefaultButton() 方法。

创建自定义对话框时，要注意对用户操作的判断，必要时需要设置响应的变量进行状态接收。例如，return this.userSaved; 返回用户操作标志。

实践 5-4 文件选择器和颜色选择器

1. 实践结果

本例运行结果如图 5-7、图 5-8 所示。

2. 实践目的

了解文件选择器和颜色选择器。

（1）文件选择器

文件选择器 JFileChooser 可以让用户浏览文件系统，便于选择一个或多个文件。该类不是 JDialog 的子类，使用时需要调用 showOpenDialog 方法或 shoSaveDialog 方法，或 showDialog 方法，无论使用哪种方法显示它，均为模态对话框。文件选择器用法比较简单，步骤如下。

创建 JFileChooser 对象：

```
JFileChooser choose = new JFileChooser();
```

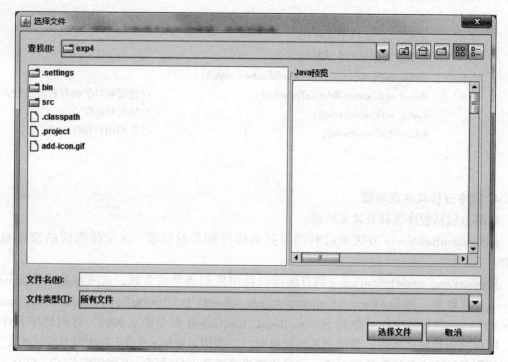

图 5-7 文件选择器

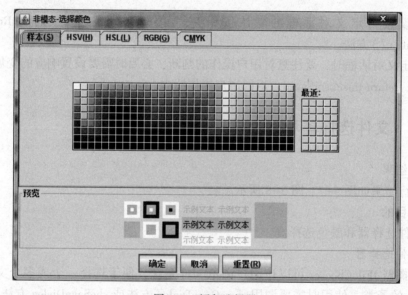

图 5-8 颜色选择器

设置工作目录：

  choose.setCurrentDirectory(new File("."));

设置默认文件：调用 setSelectedFile 方法。

根据需要添加或设置文件过滤器、文件视图，添加附加组件以增强功能：addChoosableFileFilter、setFileView、setAccessory。

设置是否可以进行文件多选：

    setMultiSelectionEnabled(true)

显示并获取返回值：

    if (choose. showDialog(this, "选择文件") == JFileChooser. APPROVE_OPTION){…}

（2）颜色选择器

  颜色选择器 JColorChooser 可以让用户方便地选择颜色，与文件选择器一样，它也不是对话框，而是一个组件。使用时，可以设为模态和非模态。

设为模态时：

    Color color = JColorChooser. showDialog(this, "模态-选择颜色",
        Color. GREEN);

设为非模态时，先创建一个颜色选择器对象：

    JColorChooser choose = new JColorChooser();

再创建非模态对话框：

```
JDialog dialog = JColorChooser. createDialog(this,  //对话框的父组件
        "非模态-选择颜色",                           //对话框标题
        false,                                       //非模态
        choose,                                      //已创建的颜色选择器
        event3 -> {                                  //OK 按钮的监听器
            label. setForeground(choose. getColor());
        }, null                                      //Cancel 按钮的监听器
);
```

最后显示：

    dialog. setVisible(true);

3. 程序源代码

程序的源代码如下：

```
package exp4;

import java. awt. BorderLayout;
import java. awt. Color;
import java. awt. Dimension;
import java. awt. EventQueue;
import java. awt. Font;
import java. awt. HeadlessException;
import java. io. BufferedReader;
import java. io. File;
import java. io. FileNotFoundException;
```

```java
import java.io.FileReader;
import java.io.IOException;

import javax.swing.BorderFactory;
import javax.swing.Icon;
import javax.swing.ImageIcon;
import javax.swing.JButton;
import javax.swing.JColorChooser;
import javax.swing.JDialog;
import javax.swing.JFileChooser;
import javax.swing.JFrame;
import javax.swing.JLabel;
import javax.swing.JMenuBar;
import javax.swing.JPanel;
import javax.swing.JScrollPane;
import javax.swing.JTextArea;
import javax.swing.filechooser.FileFilter;
import javax.swing.filechooser.FileNameExtensionFilter;
import javax.swing.filechooser.FileView;

public class Exp4 {

    public static void main(String[] args) {
        //TODO Auto-generated method stub
        EventQueue.invokeLater(() -> {
            BlankFrame frame = new BlankFrame(600, 800);         //实例化一个框架
            frame.setDefaultCloseOperation(JFrame.EXIT_ON_CLOSE);//定义关闭时的动作
            frame.setTitle("Hello JFrame!");                     //设置框架标题
            frame.setLocationByPlatform(true);                   //由窗口系统定位框架
            frame.pack();
            frame.setVisible(true);                              //显示框架
        });
    }

}

class BlankFrame extends JFrame {
    int w, h;

    private JMenuBar menuBar;

    public BlankFrame(int w, int h) throws HeadlessException {
```

```java
        super();

        this.w = w;
        this.h = h;
        this.setLayout(new BorderLayout());                    //边框布局
        JPanel panel = new JPanel();                           //增加一个Panel
        panel.setPreferredSize(new Dimension(400, 400));       //手动改变JPanel首选大小
        this.add(panel, BorderLayout.CENTER);                  //panel放中间

        JButton fileChoose = new JButton("文件对话框");        //创建一个按钮打开文件选择器
        fileChoose.addActionListener((event) -> {              //添加监听器
            JFileChooser choose = new JFileChooser();          //创建文件选择对象
            choose.setCurrentDirectory(new File("."));         //设置当前目录

            choose.setMultiSelectionEnabled(true);             //允许多文件选择

            //使用一个自定义的Java文件的文件过滤器,仅显示Java文件
            choose.addChoosableFileFilter(new JavaFileFilter());

            //添加一个图像文件的过滤器,仅显示图像
            choose.addChoosableFileFilter(new FileNameExtensionFilter(
                "图像文件", "gif", "jpg"));

            //设置文件视图,可以指定Java文件的图标
            choose.setFileView(new JavaFileView(new JavaFileFilter(),
                new ImageIcon("add-icon.gif")));

            //设置附件组件,用于显示选择的Java文件内容
            choose.setAccessory(new JavaFilePreviewer(choose, 100, 100));

            //显示文件选择器对话框,获取用户单击行为
            if (choose.showDialog(this, "选择文件") == JFileChooser.APPROVE_OPTION) {
                System.out.println("你选择了:");
                for (File f : choose.getSelectedFiles()) {
                    System.out.println(f.getPath());
                }
            }
        });

        //创建一个Label,观察文字颜色
        JLabel label = new JLabel("颜色块", JLabel.CENTER);
        //设置label的字体,大小和前景色
```

```java
        label.setFont(new Font(Font.SANS_SERIF, Font.BOLD, 50));
        label.setForeground(Color.GREEN);
        label.setPreferredSize(new Dimension(200, 200));

        //创建用于显示颜色选择器(模态)的按钮,并添加监听器
        JButton colorSelect = new JButton("颜色对话框-模态");
        colorSelect.addActionListener(event -> {
            //显示对话框,并获得选择的颜色
            Color color = JColorChooser.showDialog(this, "模态-选择颜色",
                    Color.GREEN);
            label.setForeground(color);
        });
        //创建用于显示颜色选择器(非模态)的按钮,并添加监听器
        JButton colorSelect2 = new JButton("颜色对话框-非模态");
        colorSelect2.addActionListener(event -> {
            //创建颜色选择器,并设置其初始颜色
            JColorChooser choose = new JColorChooser();
            choose.setColor(Color.GREEN);
            //对颜色选择器背后的模型,添加监听器
            choose.getSelectionModel().addChangeListener(event2 -> {
                label.setForeground(choose.getColor());
            });
            //创建非模态对话框,并显示
            JDialog dialog = JColorChooser.createDialog(this,   //对话框的父组件
                    "非模态-选择颜色",                              //对话框标题
                    false,                                      //非模态
                    choose,                                     //已创建的颜色选择器
                    event3 -> {                                 //OK按钮的监听器
                        label.setForeground(choose.getColor());
                    }, null                                     //Cancel按钮的监听器
            );
            dialog.setVisible(true);                            //显示
        });
        //添加各种组件
        this.add(label, BorderLayout.CENTER);
        this.add(fileChoose, BorderLayout.NORTH);
        this.add(colorSelect, BorderLayout.EAST);
        this.add(colorSelect2, BorderLayout.WEST);
    }
}

/**
 * @author Administrator 文件过滤器,只有被文件过滤器接受的文件才会在文件选择器中显示
```

```java
         */
        class JavaFileFilter extends FileFilter {

            public static String suffix(String filename) {          //取后缀
                int index = filename.lastIndexOf(".");
                if (index == -1) {
                    return "";
                }
                String suffix = filename.substring(index + 1);
                return suffix;
            }

            @Override
            public boolean accept(File arg0) {                       //是否接受(显示)文件

                //TODO Auto-generated method stub
                if (arg0.isDirectory()) {                            //注意文件夹的操作
                    return true;
                } else {
                    if (suffix(arg0.getName()).equals("java")) {     //判断 Java 后缀
                        return true;
                    } else {
                        return false;
                    }
                }

            }

            @Override
            public String getDescription() {                         //过滤器的文字描述
                //TODO Auto-generated method stub
                return "Java 代码文件";
            }

        }

        /**
         * @author Administrator 文件视图类,可以让文件选择器对每个文件显示特定的图标和文件描述
         */
        class JavaFileView extends FileView {

            private FileFilter filter;                               //过滤器
            private Icon icon;                                       //图标
```

```java
public JavaFileView(FileFilter filter, Icon icon) {
    this.filter = filter;
    this.icon = icon;
}

@Override
public Icon getIcon(File arg0) {                          //这里根据扩展名选择Icon

    //TODO Auto-generated method stub

    if (arg0.isDirectory()) {                             //文件夹
        return null;                                      //返回null则使用默认的视图
    } else {
        System.out.println("处理" + arg0.getName());
        if (filter.accept(arg0)) {
            return icon;
        } else {
            return null;
        }
    }
}
}

/**
 * @author Administrator 这是一个自定的组件,扩展自JPanel的目的是为文件选择器中选中的
 Java文件提供内容预览
 *
 */
class JavaFilePreviewer extends JPanel {
    JTextArea area;                                       //用于显示文件内容

    private void loadJavaFile(File f) {                   //将文件内容加载到JTextArea
        area.setText("");
        FileReader reader;
        try {
            reader = new FileReader(f);                   //定义fileReader对象
            BufferedReader buffer = new BufferedReader(reader);//BufferedReader对象将文件
                                                          //内容读取到缓存
            String s = "";
            while ((s = buffer.readLine()) != null) {     //逐行读取文件内容
                area.append("\n");
                area.append(s);
            }
```

```java
            buffer.close();
        } catch (FileNotFoundException e) {
            //TODO Auto-generated catch block
            e.printStackTrace();
        } catch (IOException e) {
            //TODO Auto-generated catch block
            e.printStackTrace();
        }
    }

    public JavaFilePreviewer(JFileChooser chooser, int arg0, int arg1) {
        super();
        //TODO Auto-generated constructor stub
        //初始化 JTextArea
        area = new JTextArea(arg0, arg1);
        setLayout(new BorderLayout());
        //给 area 加上滚动窗格
        JScrollPane scrollPane = new JScrollPane(area);
        add(scrollPane, BorderLayout.CENTER);
        //加入一个边框题目
        setBorder(BorderFactory.createTitledBorder("Java 预览"));
        //给指定的文件选择器对象,添加监听器
        chooser.addPropertyChangeListener(event -> {
            //判断文件选择器属性变化时,是否为特定属性
            if (event.getPropertyName() == JFileChooser.SELECTED_FILE_CHANGED_PROPERTY) {
                File selFile = (File) event.getNewValue();    //获取属性的新值
                //根据情况更新文本区
                if (selFile == null) {
                    area.setText("没有选择文件!");
                } else {
                    if (selFile.getName().endsWith(".java")) {
                        this.loadJavaFile(selFile);
                    }
                }
            }
        });
        this.setPreferredSize(new Dimension(300, 300));
    }
}
```

4. 程序分析与注意事项

此部分包括程序解释及常见问题。

文件过滤器需要继承父类 FileFilter，并覆盖 accept 和 getDescription 两个方法。文件选择器的 addChoosableFileFilter 方法可以向文件选择器添加多个文件过滤器。

文件视图可以为文件选择器中的每个文件提供特定的图标和描述，自定义的文件视图需要继承 FileView 父类。文件选择器的 setFileView 方法可以将文件视图安装到文件选择器。

附加组件可以是 Swing 中任意的组件，利用文件选择器的 setAccessory 方法附加在文件选择器中，增强其功能。本例中的附加组件 JavaFilePreviewer 主要用于对 Java 文件的内容进行预览。为了让附加组件及时按照用户选择的文件更新预览内容，需要向文件选择器加入属性修改监听器，例如 chooser. addPropertyChangeListener(event ->{…}}。

颜色选择器可以是非模态的，因此可以让应用程序的其他部分随着用户颜色的选择而动态更新。例如本例中 label 的前景色，随着用户选择颜色的不同，而同步发生变化，无须等到颜色选择器关闭。为了达到这个效果，需要对颜色选择器背后的模型添加监听器。

例如：

```
choose. getSelectionModel( ). addChangeListener( event2 -> {
        label. setForeground( choose. getColor( ) );
});
```

练习题

1. 编辑一个简单的通讯录。
2. 编辑一个简单的计算器，具有一定的运算功能。
3. 编辑一个简单的图像浏览器，并可用单击【上一张】和【下一张】两个按钮查看图片。

第 6 章 异常及多线程

异常是在程序运行过程中发生的错误，有可能是网络连接的问题，也有可能是无效的数组下标等。在编写程序时，必须考虑到可能发生的异常情况并做出相应的处理。异常处理的任务就是要保障用户能返回到一种较为安全的状态，并对当前的结果进行适当的保存。

多线程与多进程不同，每个进程拥有自己的一套变量而线程则共享数据，通常我们可以将多线程理解为一个程序同时执行多个任务，这些任务就是一个个线程。支持一个以上线程的应用程序称为多线程程序。

在这一章中，将向读者介绍异常及多线程的概念，异常处理的基本方法，基本的异常类，创建用户自己的异常，线程的实现及控制，多线程中的互斥，资源的同步等知识。

实践 6-1 异常的基础知识

1. 实践结果

本例的运行结果如图 6-1 所示。本例将打开一个本地文件并输出，随后创建一个数组，并输出其中的元素。

```
https://docs.oracle.com/javase/8/docs/api/
The Throwable class is the superclass of all errors and exceptions in th
Instances of two subclasses, Error and Exception, are conventionally use
A throwable contains a snapshot of the execution stack of its thread at
One reason that a throwable may have a cause is that the class that thro
A second reason that a throwable may have a cause is that the method tha
A cause can be associated with a throwable in two ways: via a constructo
By convention, class Throwable and its subclasses have two constructors,
0
1
2
3
4
5
Exception in thread "main" java.lang.ArrayIndexOutOfBoundsException: 6
        at exp1.Exp1.main(Exp1.java:23)
```

图 6-1 实践 6-1 的控制台输出

2. 实践目的

本例介绍异常的基本概念，异常的处理机制，异常类，异常类的方法和域。

（1）异常的基本概念

异常是在程序运行过程中发生的、会打断程序正常执行的事件。比如打开文件失败、除数为 0、数组元素下标越界等。本例在打开文件时没有遇到问题（readme.txt 存在），而在随后的数组访问时，由于存在下标越界的情况，因此执行时会遇到一个下标越界异常。

(2) Java 对异常的处理机制

Java 使用异常对 Java 程序给出了一个统一和相对简单的抛出和处理错误的机制。如果一个方法本身抛出异常（利用 throw 和 throws），调用者可以捕捉异常使之得到处理；也可以回避异常，这时异常将在调用的堆栈中向下传递，直到被处理。在堆栈底部的方法必须捕获和处理所有还没处理的异常，否则程序就会出问题。

(3) 抛出一个异常

在 Java 程序的运行过程中，如果某个方法执行时产生了异常，则这个方法代表该异常的一个异常对象（也可能由 Java 虚拟机 JVM 产生），并把它交给"运行时系统"，"运行时系统"将寻找相应的代码来处理这个异常，生成异常对象并把它提交给"运行时系统"的过程称为抛出（throw）一个异常。如图 6-1 所示，已经在 Exp1.java 的第 23 行捕获了一个异常。

(4) 捕获一个异常

"运行时系统"得到一个异常对象时，将寻找处理该异常的代码。通过方法的调用栈查找，从生成异常的方法开始进行回溯，直到找到包含相应异常处理的方法为止。然后"运行时系统"把该异常对象交给这个方法进行处理，这个过程称为捕获一个异常。捕获异常采用 try{}catch(){}{}代码块，本例代码中打开文件 readme.txt 的语句被包含在 try{}中，而相应的异常处理放在 catch 代码块中，finally 代码块中的代码无论出现异常与否总会被执行，如图 6-2 所示。

```
Scanner scan = null;
try {// 打开文件，输出文件内容
    scan = new Scanner(new File("readme.txt"));
    while (scan.hasNext()) {
        System.out.println(scan.nextLine());
    }
} catch (FileNotFoundException e) {
    // TODO Auto-generated catch block
    e.printStackTrace();
} finally {//无论是否出现异常都会执行的部分
    if (scan != null)
        scan.close();//关闭操作
}
```

图 6-2 捕获异常代码块示例

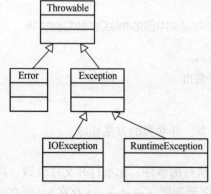

图 6-3 异常处理的类层次图

(5) 异常类层次结构

Java 用面向对象的方法处理异常，一个异常事件由一个异常类的对象来代表。Java 语言的异常处理的类层次图如图 6-3 所示。

所有异常都是由 Throwable 继承而来的，其子类为 Error 和 Exception。Error 类分支主要用于描述系统内部错误和资源枯竭错误，因此普通的应用程序不应抛出此类异常。Exception 类分支主要分为 IOException 和 RuntimeException。RuntimeException 属于由程序本身引发的异常，即程序本身存在错误；IOException 则用于其他不属于程序本身错误所引发的异常。

(6) 异常使用规则

遇到 RuntimeException 则表明,程序本身出现了问题,例如本例中的数组遍历,本应当进行下标检查,却没有检查从而导致的下标越界异常。这类问题是可以避免的。

区分环境问题和代码问题,本例对文件打开操作使用了 try..catch 的异常捕获和处理代码,这是因为文件的状态属于外部因素,这一点并非程序可以控制的,因此不是代码问题,而是环境问题。

Java 将 Error 类和 RuntimeException 类的分支定义为"非受查异常",即编译器并不会检查是否为它们提供了异常处理器。相反,其他异常类型则属于"受查异常",编译器会检查它们是否具有异常处理器。本例中 Scanner 的构造函数会抛出一个 FileNotFoundException 异常,它是 IOException 的子类,因此编译器会强制检查,并要求提供异常处理器。

3. 程序源代码

程序的源代码如下:

```java
package exp1;

import java.io.File;
import java.io.FileNotFoundException;
import java.io.IOException;
import java.util.Scanner;

public class Exp1 {

    public static void main(String[] args) {
        //TODO Auto-generated method stub
        Scanner scan = null;
        try {//打开文件,输出文件内容
            scan = new Scanner(new File("readme.txt"));
            while (scan.hasNext()) {
                System.out.println(scan.nextLine());
            }
        } catch (FileNotFoundException e) {
            //TODO Auto-generated catch block
            e.printStackTrace();
        } finally {//无论是否出现异常都会执行的部分
            if (scan != null)
                scan.close();//关闭操作
        }
        //创建一个数组并访问
        int[] a = {0, 1, 2, 3, 4, 5};
        for (int i = 0; i < 10; i++) {
            System.out.println(a[i]);
```

```
            }
        System.exit(0);
    }
}
```

4. 程序分析与注意事项

此部分包括程序解释及常见问题。

派生于 RuntimeException 的异常主要包括：类型转换错误、数组下标越界、Null 指针等。

并非派生 RuntimeException 的异常主要有：文件读取错误、文件不存在、查找不存在等。

printStackTrace( )可以访问堆栈轨迹的文本描述信息。

对于受查异常，程序员必须利用 try...catch 代码块进行捕获并提供异常处理器，否则无法通过编译。通常集成开发环境会给出提示信息，并辅助生成代码，如图 6-4 所示。

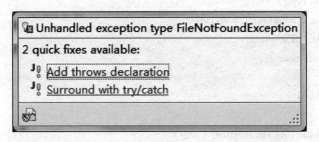

图 6-4　出现未捕获的受查异常时 Clipse 的提示信息

实践 6-2　自定义异常

1. 实践结果

本例的运行结果如图 6-5 所示。

```
exp2.FileHeaderException: 文件开头有问题:文件头为https://docs.oracle.com/javase/8/docs/api/
    at exp2.ZWFileReader.checkHeader(Exp2.java:62)
    at exp2.Exp2.main(Exp2.java:15)
```

图 6-5　实践 6-2 的控制台输出

2. 实践目的

本例主要是从自定义异常入手，介绍异常处理中常用的几个技巧。

创建用户自定义异常的语法格式如下：

　　class　自定义异常　extends　父异常类名{类体;}

（1）用户自定义类的类头通常继承自 Exception 类或 Exception 类的子类。

（2）通常自定义异常类需要提供一个空的构造函数和一个带有详细信息的构造函数。

（3）在满足一定条件下，通过 throw 语句抛出用户自定义异常的对象。抛出用户自定义异常的语法格式如下：

```
［修饰符］ 类型  方法名(参数1,参数2,…)  throws 自定义异常
{
    …
    if(条件满足)
        throw(new myException(this));
    else
        …
}
```

3. 程序源代码

程序的源代码如下：

```java
package exp2;

import java.io.File;
import java.io.FileInputStream;
import java.io.FileNotFoundException;
import java.io.IOException;
import java.util.Scanner;

public class Exp2 {

    public static void main(String[] args) {
        //TODO Auto-generated method stub
        ZWFileReader fileReader=new ZWFileReader("readme.zw");
        try {
            fileReader.checkHeader();
        } catch (FileNotFoundException | FileHeaderException e) {
            //可以在一个catch中捕获多个异常，前提条件是这两个或多个异常不存在父子关系
            //且此时 e 隐含为 final，不可修改它，这一点与分开写两个 catch 不同
            e.printStackTrace();
        }
    }

    /**
     * @author Administrator 自定义异常，用于表述文件开头存在问题的错误。因此不应该从 Error
     * 和 RuntimExcption 继承
```

```
 *   考虑从 Exception 或 IOException 继承均可。
 */
class FileHeaderException extends IOException {

    public FileHeaderException() {
        super();
    }

    public FileHeaderException(String arg0) {
        super("文件开头有问题:" + arg0);
    }

}


/**
 * @author Administrator
 * 自定义一个文件读入类,检查文件第一行
 * 当不满足要求时,抛出自定义异常。
 */
class ZWFileReader {
    private String fileName;

    public ZWFileReader(String fileName) {
        this.fileName = fileName;
    }

    public void checkHeader() throws FileNotFoundException, FileHeaderException {
        String line;
        try (Scanner scan = new Scanner(new File(this.fileName))) {
            //如果资源的类型实现了 AutoClose 接口,则可以放入 try()中
            //当 try 退出时,会自动调用接口中的 close 方法,从而实现自动的关闭。无论是否
            //出现异常。
            //这样减少了 finally 的编写,代码更加整洁
            line = scan.nextLine();
            if (!line.equals("ZW")) {
                //抛出一个自定义异常
                throw new FileHeaderException("文件头为" + line);
            }
        }
    }
}
```

4. 程序分析与注意事项

在本例中定义了一个自定义异常类和一个文件读取类,checkHeader()方法检查文件的

第一行，当内容不满足时抛出自定义异常。

（1）catch( )中可以捕获多个异常

例如 catch（FileNotFoundException | FileHeaderException e）{ }，前提条件是这两个或多个异常不存在父子关系，而且此时 e 隐含为 final 变量，异常处理中不可修改它，这一点与普通的 catch 语句块不同。

（2）自定义异常的父类选择

通常自定义异常的父类不应选择 Error 类分支，因为 Error 类分支代表系统内部错误和资源枯竭错误，这类错误用户是无能为力的。

（3）带有资源的 try 语句

当 try 语句中需要使用资源，而且这个资源需要在 finally 中关闭时，考虑使用带资源的 try 语句。例如 try（Scanner scan = new Scanner(new File(this.fileName)))｛…｝，前提条件是这个资源类实现了 AutoCloseable 接口。带有资源的 try 语句会自动关闭资源，即调用 Auto-Closeabel 接口中的 close 方法，而无论是否发生了异常。

（4）异常抛出的原则

通常捕获知道如何处理的异常；对于不知如何处理的异常，则让它们继续传递下去，例如 checkHeader 方法，并未捕获任何异常，而是让它继续传递下去。即让 checkHeader 方法的调用者去决定，到底该怎么处理。

实践 6-3 多线程入门

1. 实践结果

本例的运行结果如图 6-6 所示。本例中可以单击【开始】按钮向面板中增加一个球（可以多次单击，从而添加多个球），之后球开始不断的运动；单击【退出】按钮后，结束程序。

2. 实践目的

本例介绍多线程的概念、实现方法和基本控制。

（1）多线程就是在一个程序内部同时进行多个任务，每个任务占用一个线程。由于线程在程序内部，多个线程共享一些变量，这比在进程间的通信更加有效和容易。因此线程更加"轻量级"，创建和撤销一个线程比启动进程开销要小。

（2）创建线程的基本方法

创建一个实现 Runnable 接口的类，实现接口中的 run 方法，将任务代码放置在 run 方

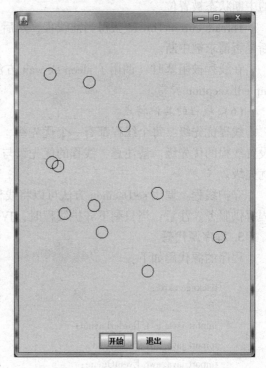

图 6-6 实践 6-3 的运行界面

法体内。

由 Runnable 创建一个 Thread 对象，并调用 start 方法启动线程。

(3) Runnable 接口

Runnable 接口中只有一个方法 run，因此属于函数式接口，可以利用 lambda 表达式建立接口的实例。从而省去重新定义一个类的烦琐步骤。

(4) 线程状态

可以利用 getState 方法获得线程的状态，包含以下 6 种。

New（新创建）是指用 new 操作符刚刚创建了线程对象，还没有开始运行。

Runnable（可运行）是指一旦调用了线程的 start 方法，则线程处于可运行状态，但注意此时线程可能正在运行，也可能没有运行，这一切取决于操作系统。

Blocked（被阻塞）是指线程试图获得一个内部的对象锁（但不是 java.util.concurrent 中的锁），而该锁被其他线程持有时，线程进入阻塞状态。此时线程暂时没有活动，直到线程调度器再次激活它。

Waiting（等待）是指线程等待另一个线程通知线程调度器一个条件时，线程进入等待状态，此时它暂时没有活动，直到线程调度器再次激活它。

Timed waiting（计时等待）是指在等待基础上，增加了一个超时参数。

Terminated（被终止）是指线程结束，通常线程会在 run 方法执行退出后死亡，或者因为一个未捕获异常而终止。

(5) 中断线程

没有可以强制线程终止的方法，但可以调用线程的 interrupt 方法请求它终止。之后线程的中断状态被置位。

Thread.currentThread 获得当前线程，然后调用 isInterrupted 方法检测中断状态，从而判断是否请求被中断。

在线程被阻塞时（调用了 sleep 或 wait 方法），如果调用 interrupt 方法，则会引发 InterruptedException 异常。

(6) 线程的其他特点

线程优先级：每个线程都有一个优先级，默认继承自父线程。调用 setPriority 方法可以设置线程的优先级。请注意，线程的优先级与系统有关，因此不要将程序的正确执行依赖于优先级。

守护线程：调用 setDaemon 方法可以将线程设置为守护线程，它的主要作用是为其他线程提供服务。注意：当只剩下守护线程时，JVM 会退出。

3. 程序源代码

程序的源代码如下：

```
package exp3;

import java.awt.BorderLayout;
import java.awt.Dimension;
import java.awt.EventQueue;
import java.awt.Graphics;
```

```java
import java.awt.Graphics2D;
import java.awt.HeadlessException;
import java.awt.geom.Ellipse2D;
import java.awt.geom.Ellipse2D.Double;
import java.util.ArrayList;
import java.util.Collections;
import java.util.List;
import java.util.Random;

import javax.swing.JButton;
import javax.swing.JFrame;
import javax.swing.JPanel;

public class Exp3 {

    public static void main(String[] args) {
        //TODO Auto-generated method stub
        EventQueue.invokeLater(() -> {//将 Swing 代码放置在事件分配线程中
            JFrame frame = new SimpleFrame();
            frame.setDefaultCloseOperation(JFrame.EXIT_ON_CLOSE);
            frame.pack();
            frame.setVisible(true);
        });
    }
}

class SimpleFrame extends JFrame {

    private DrawBallComponent drawBallComponent;    //绘制小球的自定义组件

    public SimpleFrame() throws HeadlessException {
        super();

        drawBallComponent = new DrawBallComponent();
        this.add(drawBallComponent, BorderLayout.CENTER);//绘制小球的自定义组件放置在中间

        JPanel buttonsPanel = new JPanel();           //按钮面板
        JButton addButton = (JButton) buttonsPanel.add(new JButton("开始"));
        JButton exitButton = (JButton) buttonsPanel.add(new JButton("退出"));

        addButton.addActionListener((event) -> {    //"开始"按钮的监听器
            MyBall ball = drawBallComponent.addBall();   //增加一个球
            //并把这个球的移动放置在一个线程中
```

```java
            //Runnable 接口只有一个方法,是一个函数式接口,可以用 lambda 表达式建立
            //实例
            Runnable r = () -> {
                for (int i = 0; i < 1000; i++) {

                    ball.move(DrawBallComponent.width,
                            DrawBallComponent.height);

                    drawBallComponent.repaint();
                    try {
                        Thread.sleep(10);
                    } catch (Exception e) {
                        //TODO Auto-generated catch block
                        e.printStackTrace();
                    }

                }
            };
            //利用 Runnable 变量创建一个线程,并启动这个线程
            Thread t = new Thread(r);
            t.start();
        });
        exitButton.addActionListener((event) -> {
            System.exit(0);
        });
        this.add(buttonsPanel, BorderLayout.SOUTH);//按钮面板放在底部
    }

}

/**
 * @author Administrator 球类
 */
class MyBall {
    static double W = 20;        //椭圆的宽
    static double H = 20;        //椭圆的高
    private double leftX, topY;

    public MyBall(double leftX, double topY) {
        super();
        this.leftX = leftX;
        this.topY = topY;
    }
```

```java
/**
 * 随机计算球的位置,模拟移动
 *
 * @param maxX
 *              X 坐标边界
 * @param maxY
 *              Y 坐标边界
 */
public void move(double maxX, double maxY) {
    this.leftX = (this.leftX * (1 + Math.random())) % maxX;
    this.topY = (this.topY * (1 + Math.random())) % maxY;
}


/**
 * 获得绘图需要的椭圆形
 *
 * @return
 */
public Ellipse2D.Double getShape() {
    //宽和高一致来制作圆
    Ellipse2D.Double shape = new Ellipse2D.Double(leftX, topY, W, H);
    return shape;


    }
}


class DrawBallComponent extends JPanel {
    static int width = 400;
    static int height = 500;
    private List<MyBall> balls = Collections
            .synchronizedList(new ArrayList<MyBall>());

    //ArrayList 和 HashMap 都不是线程安全的。
    //使用同步包装器,将集合类变成线程安全的。表明该代码可以被多线程使用,线程的调度
    //顺序不影响任何结果。
    //但是在遍历时(如果此时存在被修改的可能)应加上锁,否则会抛出 ConcurrentModification
    //Exception
    @Override
    protected void paintComponent(Graphics arg0) {
        //TODO Auto-generated method stub
        super.paintComponent(arg0);             //清除背景
        Graphics2D g = (Graphics2D) arg0;
        drawAllBall(g);
```

```java
    }

    private void drawAllBall(Graphics2D g){
        synchronized(balls){          //一个线程要想进入这个代码块必须先获得balls的锁
            //遍历列表进行绘图
            for(MyBall ball : balls){
                g.draw(ball.getShape());
            }
        }
    }

    public MyBall addBall(){          //加入一个球
        MyBall ball = new MyBall(Math.random() * width, height * Math.random());
        synchronized(balls){          //一个线程要想进入这个代码块必须先获得balls的锁
            this.balls.add(ball);
        }

        return ball;
    }

    @Override
    public Dimension getPreferredSize(){
        //TODO Auto-generated method stub
        return new Dimension(width, height);
    }

}
```

4. 程序分析与注意事项

此部分包括程序解释及常见问题。

Swing 不是线程安全的，因此一个耗时较长的任务需要放在单独的工作线程中，而不要放在事件分配线程中。例如，给每一个球的运动单独启动一个线程，而不是放在事件分配线程中。就像本例中"开始"按钮的监听器那样：

```java
addButton.addActionListener((event)->{//开始按钮的监听器
    MyBall ball = drawBallComponent.addBall();//增加一个球
    //并把这个球的移动放置在一个线程中
    //Runnable 接口只有一个方法，是一个函数式接口，可以用 lambda 表达式建立实例
    Runnable r=()->{
        for(int i = 0; i < 1000; i++){

            ball.move(DrawBallComponent.width,
```

```
                    DrawBallComponent. height);

                drawBallComponent. repaint( );
                try {
                    Thread. sleep(10);
                } catch (Exception e) {
                    //TODO Auto-generated catch block
                    e. printStackTrace( );
                }
            }
        };
        //利用 Runnable 变量创建一个线程，并启动这个线程
    Thread t = new Thread(r);
    t. start( );
});
```

ArrayList 和 HashMap 都不是线程安全的，这意味着当它们被多线程使用时，线程的调度顺序影响最终的结果。因此需要使用同步包装器对球列表 balls 进行包装，如：

```
private List<MyBall> balls = Collections. synchronizedList( new ArrayList<MyBall>( ) )
```

每一个 Java 对象有一个锁。由于在 addBall 方法和 drawAllBall 方法中均对 balls 的内容进行了访问（前者是增加一个球，后者是遍历所有的球），因此在两个方法内部使用 synchronized 获得 balls 的锁之后才能进行遍历和增加球的操作，这种利用一个对象的锁实现原子操作的方法，也称为"客户端锁"。

此外，当一个方法用 synchronized 关键字声明后，当前 Java 对象的内部锁将保护这个方法。一个线程要想调用这个方法，必须获得对象的内部锁。

实践 6-4 多线程的同步

1. 实践结果

本例的运行结果如图 6-7 所示。

```
Thread[Thread-61,5,main]Account38->Account94 =34 total=10000
Thread[Thread-32,5,main]Account29->Account45 =69 total=10000
Thread[Thread-59,5,main]Account45->Account48 =69 total=10000
Thread[Thread-17,5,main]Account65->Account85 =99 total=10000
Thread[Thread-5,5,main]Account85->Account32 =69 total=10000
Thread[Thread-32,5,main]Account25->Account52 =60 total=10000
Thread[Thread-49,5,main]Account52->Account8 =64 total=10000
Thread[Thread-99,5,main]Account29->Account88 =29 total=10000
Thread[Thread-83,5,main]Account69->Account9 =27 total=10000
Thread[Thread-82,5,main]Account31->Account68 =65 total=10000
Thread[Thread-7,5,main]Account62->Account65 =73 total=10000
```

图 6-7 实践 6-4 的控制台输出

2. 实践目的

本例将向读者介绍多线程的同步问题。我们在编写程序时会建立多个线程,这些线程共享一些资源对象,如:银行在各账户间转移数据,各账户数据就是共享资源对象;在扑克游戏中各个玩家牌的不断操作,这些牌也是共享资源对象。各个线程之间经常轮流挂起、恢复运行来操作这些共享资源对象,如果不协调这些操作的话,就不能保证运行结果的正确。

本例模拟构建了一个银行,开启了100个线程对银行中的账户进行随机转账,同时需要保障银行中所有账户的金额之和,保持不变。

在 Java 中提供了两种机制来防止代码块受到并发线程的影响。synchronized 关键字是其中之一,另外一个机制是使用 java.util.concurrent 中提供的锁对象和条件对象,本例使用 ReentrantLock 类创建了一个锁对象,并使用锁对象的 newCondition 方法创建了一个与该锁相关的条件对象。

① 锁对象能够保证在同一时刻,只有一个线程进入代码块并锁住锁对象,此后其他任何线程都无法通过锁对象的 lock 方法,试图调用 lock 方法的线程会被阻塞。锁对象使用的基本机构如下:

```
aLock.lock();
try{ ... }
finally{...alock.unlock();}
```

对于锁对象的 unlock() 方法务必置于 finally 代码块中,否则其他线程会因为无法获得锁而被阻塞。

② 一个锁对象可以拥有多个相关的条件对象。条件对象调用 await() 方法会使当前线程阻塞并放弃锁。当前线程会一直处于该条件的等待状态,直到有其他线程调用了同一个条件对象的 signallAll 方法为止。通常应当循环对条件进行检测,只要条件不允许,则调用 await() 方法进入阻塞。基本格式如下:

```
while(条件不具备检测){
    condition.await();
}
```

3. 程序源代码

程序的源代码如下:

```
package exp4;

import java.util.ArrayList;
import java.util.List;
import java.util.concurrent.TimeUnit;
import java.util.concurrent.locks.Condition;
import java.util.concurrent.locks.Lock;
import java.util.concurrent.locks.ReentrantLock;

public class Exp4 {
```

```java
public static void main(String[] args) {
    //TODO Auto-generated method stub
    MyBank bank = new MyBank(100, 100);
    for (int i = 0; i < 100; i++) {
        //启动100个线程,每个线程在不停地进行随机转账的操作
        Runnable r = () -> {
            while (true) {
                Account from, to;
                from = bank.accounts
                        .get((int) (Math.random() * bank.accounts.size()));
                to = bank.accounts.get((int) (Math.random() * bank.accounts
                        .size()));
                //bank.transUnsynch(from, to, (int) (100 * Math.random()));
                //非同步,会导致总账户金额不为10000

                //bank.transSynch(from, to, (int) (100 * Math.random()));
                //同步,没有使用条件对象,金额不够则取消

                bank.transSynchWithContition(from, to,
                        (int) (100 * Math.random()));
                //同步,并使用条件对象,当条件不满足时让线程阻塞,如果不使用等待
                //超时参数则会引发死锁
                //等待一段时间,条件仍不满足则取消
                try {
                    Thread.sleep((int) (Math.random() * 100));
                } catch (InterruptedException e) {
                    //TODO Auto-generated catch block
                    e.printStackTrace();
                }
            }
        };
        Thread thread = new Thread(r);
        thread.start();
    }

}

/**
 * @author Administrator
 * 一个模拟的银行类,维护着一个账户列表
 */
```

```java
class MyBank {
    List<Account> accounts;
    int totalBalance;
    private Lock mylock;                          //使用锁对象
    private Condition enoughCash;                 //与锁对象相关的一个表示"现金充足"的条件对象

    public MyBank(int nAccount, int initAccountBalance) {
        accounts = new ArrayList<Account>(nAccount);         //账户列表
        totalBalance = nAccount * initAccountBalance;        //总账户金额
        for (int i = 0; i < nAccount; i++) {                 //初始化账户列表
            accounts.add(new Account("Account" + i, initAccountBalance));
        }
        mylock = new ReentrantLock();                        //初始化一个锁对象
        enoughCash = mylock.newCondition();                  //初始化锁上的一个条件对象
    }

    /** 非同步方法,调用这个方法会导致所有账户的总金额发生变化
     *     因为转账的操作不是原子性的,会被打断
     * @param fromAccount 转出账户
     * @param toAccount 转入账户
     * @param transCash 转账金额
     */
    public void transUnsynch(Account fromAccount, Account toAccount,
            int transCash) {
        if (fromAccount.balance >= transCash) {
            fromAccount.balance = fromAccount.balance - transCash;
            toAccount.balance = toAccount.balance + transCash;
            System.out.println(Thread.currentThread() + fromAccount.name + "->"
                    + toAccount.name + " =" + transCash + " total="
                    + getCurrentTotalBalance());
        }
    }

    /** 同步方法,没有使用条件对象,金额不够则取消
     * @param fromAccount 转出账户
     * @param toAccount 转入账户
     * @param transCash 转账金额
     */
    public void transSynch(Account fromAccount, Account toAccount, int transCash) {
        mylock.lock();        //上锁,确保只有一个线程能够进入该区域
        //一旦一个线程对 mylock 对象上锁后,其他线程都会因调用 lock 方法而被阻塞
        try {
            if (fromAccount.balance >= transCash) {
```

第 6 章 异常及多线程

```java
            fromAccount.balance = fromAccount.balance - transCash;
            toAccount.balance = toAccount.balance + transCash;
            System.out.println(Thread.currentThread() + fromAccount.name
                    + "->" + toAccount.name + " =" + transCash + " total="
                    + getCurrentTotalBalance());
        }
    } finally {
        mylock.unlock();        //解锁操作一定要放在 finally 中,这样无论是否遇到异常,
                                //都可以释放锁
    }

}

/**
 * 同步方法,并使用了条件对象,当条件不满足时让线程阻塞,如果不使用等待超时参数
 则会引发死锁
 *      等待一段时间,条件仍不满足则取消
 * @param fromAccount
 * @param toAccount
 * @param transCash
 */
public void transSynchWithContition(Account fromAccount, Account toAccount,
        int transCash) {
    mylock.lock();      //上锁,确保只有一个线程能够进入该区域
    //一旦一个线程对 mylock 对象上锁后,其他线程都会因调用 lock 方法而被阻塞
    try {
        boolean notOvertime = true;     //非超时标志
        while (fromAccount.balance < transCash) {
            notOvertime = enoughCash.await(100, TimeUnit.MILLISECONDS);//如果超时了,
                                                                        //则返回 false
            //条件对象调用 await,当前线程进入 enoughCash 条件对象的等待集,线程被阻
            //塞,并放弃了锁对象 mylock 的锁状态
            //这样才能让其他线程有机会增加账户,等待其他线程在 enoughCash 条件对
            //象上的 signalAll 方法调用
            if (!notOvertime) {             //表明超时了
                break;                      //不要再循环等待
            }
        }
        if (notOvertime) {                  //并非超时,则更新账户
            fromAccount.balance = fromAccount.balance - transCash;
            toAccount.balance = toAccount.balance + transCash;
            System.out.println(Thread.currentThread() + fromAccount.name
                    + "->" + toAccount.name + " = " + transCash + " total="
```

```java
                    + getCurrentTotalBalance());
                enoughCash.signalAll();        //通知所有等待线程
            }
        } catch (InterruptedException e) {
            //TODO Auto-generated catch block
            e.printStackTrace();
            return;
        } finally {
            mylock.unlock();    //解锁操作一定要放在 finally 中,这样无论是否遇到异常,
                                //都可以释放锁
        }
    }

    public int getCurrentTotalBalance() {
        int sum = 0;
        for (Account a : accounts) {
            sum += a.balance;
        }
        return sum;
    }
}

/**
 * @author Administrator
 * 简单的账户类
 */
class Account {
    String name;
    int balance;

    public Account(String name, int balance) {
        this.name = name;
        this.balance = balance;
    }

    @Override
    public String toString() {
        //TODO Auto-generated method stub
        return name + "的账户余额=" + balance;
    }
}
```

4. 程序分析与注意事项

此部分包括程序解释及常见问题。

(1) 非同步转账方法 transUnsynch

transUnsynch 方法会产生所有账户的金额之和变动的情况，如图 6-8 所示。

```
Thread[Thread-0,5,main]Account12->Account69 =0 total=10128
Thread[Thread-1,5,main]Account14->Account20 =48 total=10128
Thread[Thread-48,5,main]Account42->Account26 =62 total=10128
```

图 6-8 使用 transUnsynch 方法的程序输出（部分）

这主要是由于执行线程在转账过程中被打断所致，即转账操作不具备原子性，因此考虑使用锁对象。

(2) 使用锁对象的转账方法 transSynch

该方法使用了一个锁对象 mylock，用于控制该代码块的并发性，代码如下。

```
mylock.lock();        //上锁，确保只有一个线程能够进入该区域
    try {
        if (fromAccount.balance >= transCash) {
            fromAccount.balance = fromAccount.balance - transCash;
            toAccount.balance = toAccount.balance + transCash;
            System.out.println(Thread.currentThread() + fromAccount.name
                + "->" + toAccount.name + " =" + transCash + " total="
                + getCurrentTotalBalance());
        }
    } finally {
        mylock.unlock();
    }
```

一个线程对 mylock 对象上锁后，其他线程都会因调用 lock 方法而被阻塞。此外解锁操作一定要放在 finally 中，这样无论是否遇到异常，都可以释放锁。

(3) 使用锁对象和条件的转账方法 transSynchWithContition

考虑到账户余额不足时，让当前线程稍微等待，期待其他线程向该账户转账的情况，本方法使用了一个条件对象。线程循环判断转账条件，不满足时调用 await() 方法阻塞当前线程，并设置了超时参数，并释放锁。从而让其他线程有机会增加账户，并等待其他线程在 enoughCash 条件对象上的 signalAll 方法调用。

```
mylock.lock();
    try {
        boolean notOvertime = true;        //非超时标志
        while (fromAccount.balance < transCash) {
            notOvertime = enoughCash.await(100, TimeUnit.MILLISECONDS);    //如果超时了，
                                                                          //则返回 false
```

```
                    if (!notOvertime) {
                        break;                  //表明超时了
                    }                            //不要再循环等待
                }
                if( notOvertime) {              //并非超时,则更新账户
                    fromAccount. balance = fromAccount. balance - transCash;
                    toAccount. balance = toAccount. balance + transCash;
                    System. out. println( Thread. currentThread( ) + fromAccount. name
                            + "->" + toAccount. name + " =" + transCash + " total ="
                            + getCurrentTotalBalance( ));
                    enoughCash. signalAll( );    //通知本条件的所有等待线程
                }
            } catch (InterruptedException e) {
                //TODO Auto-generated catch block
                e. printStackTrace( );
                return;
            } finally {
                mylock. unlock( )
            }
```

需要说明的是,如果在 await 中不加入超时参数,则极易导致程序运行一段时间后出现死锁的情况,此时所有线程都在等待更多的钱存入其转出账户而被阻塞。Java 语言中没有可以保证避免死锁的机制,这就要求程序员仔细设计代码结构。

实践 6-5　阻塞队列

1. 实践结果

本例仍然是模拟银行转账的过程,其运行结果如图 6-9 和图 6-10 所示,主要展示在 Java 中阻塞队列的使用方法。

图 6-9　实践 6-5 的运行界面

① 对于多线程问题,很多时候可以转化为生产-消费的问题,生产者生产数据而消费者使用数据。利用阻塞队列可以安全、简洁地将数据从一个线程传递到另一个线程。

② java. util. concurrent 包中提供了若干种类的阻塞队列,本例仅以其中的 ArrayBlockQueue 进行展示,其他类型的阻塞队列请读者查阅 API 文档。

③ 本例创建了若干个线程生成转账指令,并加入到阻塞队列;同时创建一个线程读取

```
十一月 30, 2018 11:20:59 下午 exp5.MyBank trans
信息: Account11->Account19 =56 total=10000
十一月 30, 2018 11:20:59 下午 exp5.MyBank trans
信息: Account36->Account51 =47 total=10000
十一月 30, 2018 11:20:59 下午 exp5.MyBank trans
信息: Account94->Account12 =36 total=10000
十一月 30, 2018 11:20:59 下午 exp5.MyBank trans
信息: Account58->Account97 =77 total=10000
```

图 6-10　实践 6-5 的输出

阻塞队列中的转账指令完成转账的具体操作。由于银行账户信息没有被多线程共同使用，因此在银行内部的各个方法中，均没有采用同步机制。

④ 阻塞队列本身已经采用一系列线程安全的措施，因此开发人员可以放心使用。当线程试图向满队列添加元素，或是试图从空队列移出元素时，都会导致线程阻塞。此外阻塞队列通常具有自动的负载机制。

2. 程序源代码

程序的源代码如下：

```java
package exp5;

import java.awt.BorderLayout;
import java.awt.Dimension;
import java.awt.EventQueue;
import java.awt.HeadlessException;
import java.io.IOException;
import java.util.ArrayList;
import java.util.List;
import java.util.concurrent.ArrayBlockingQueue;
import java.util.concurrent.BlockingQueue;
import java.util.logging.FileHandler;
import java.util.logging.Logger;

import javax.swing.JButton;
import javax.swing.JFrame;
import javax.swing.JLabel;
import javax.swing.JPanel;
import javax.swing.JProgressBar;
public class Exp5 {

    public static void main(String[] args) {
        //TODO Auto-generated method stub
        EventQueue.invokeLater(() -> {              //将 Swing 代码放置在事件分配线程中
            JFrame frame = new SimpleFrame();
```

```java
                    frame.setDefaultCloseOperation(JFrame.EXIT_ON_CLOSE);
                    frame.pack();
                    frame.setVisible(true);
            });
    }
}

class SimpleFrame extends JFrame {
    private ShowStatusJPanel statusPanel;              //绘制小球的自定义组件
    private int value = 100;
    private MyBank bank;

    public SimpleFrame() throws HeadlessException {
        super();

        statusPanel = new ShowStatusJPanel();
        this.add(statusPanel, BorderLayout.CENTER);    //自定义组件放置在中间

        JPanel buttonsPanel = new JPanel();            //按钮面板
        JButton addButton = (JButton) buttonsPanel.add(new JButton("开始"));
        JButton exitButton = (JButton) buttonsPanel.add(new JButton("退出"));

        bank = new MyBank(100, 100);                   //创建一个银行对象

        addButton
                .addActionListener((event) -> {        //"开始"按钮的监听器

                    //启动100个线程,向阻塞队列中放置转账操作对象
                    for (int i = 0; i < 100; i++) {    //由阻塞队列进行同步问题处理
                        Runnable produce = () -> {
                            while (true) {
                                int from, to;
                                from = (int) (Math.random() * bank.accounts.size());
                                to = (int) (Math.random() * bank.accounts.size());
                                int transCount = (int) (100 * Math.random());
                                    TransInstruction tAction = new TransInstruction(from, to,
            transCount);
                                try {
                                    bank.queue.put(tAction);
                                } catch (Exception e) {
                                    //TODO Auto-generated catch block
```

```
                    e.printStackTrace();
                }
            }
        };
        Thread threadProduce = new Thread(produce);
        threadProduce.start();


//启动1个线程,从阻塞队列中取转账对象,并实施转账
//由于只有一个线程访问转账代码,所以不需要同步处理
        Runnable consume = () -> {
            while (true) {
                TransInstruction ta;
                try {
                    ta = bank.queue.take();
                    bank.trans(ta);
                } catch (Exception e) {
                    //TODO Auto-generated catch block
                    e.printStackTrace();
                }
            }

        };
        Thread threadConsume = new Thread(consume);
        threadConsume.start();

//更新控件使用的线程
        Runnable r = () -> {
            while (true) {                    //这个更新操作不停进行
                //注意:任何对于Swing控件的操作务必放置在事件队列中!!!!
                EventQueue.invokeLater(() -> {
                    value = bank.queue.size();    //获取阻塞对列中的对象数目

                    statusPanel.setValue(value);
                });
                try {
                    Thread.sleep(500);    //不用更新太频繁
                } catch (Exception e) {
                    //TODO Auto-generated catch block
                    e.printStackTrace();
                }
            }
        };
```

```java
                    Thread t = new Thread(r);
                    t.start();
                });
            exitButton.addActionListener((event) -> {
                System.exit(0);
            });
            this.add(buttonsPanel, BorderLayout.SOUTH);    //按钮面板放在底部
    }
}

/**
 * @author Administrator
 * 显示了一个JProgressBar,用于显示阻塞队列当前的状态
 */
class ShowStatusJPanel extends JPanel {
    static int width = 400;
    static int height = 50;
    private JProgressBar pbar;

    public ShowStatusJPanel() {
        super();
        pbar = new JProgressBar(0, 30);
        pbar.setValue(10);
        this.add(new JLabel("阻塞队列状态:"));
        this.add(pbar);
    }

    public void setValue(int value) {
        pbar.setValue(value);
    }

    @Override
    public Dimension getPreferredSize() {        //首选大小
        //TODO Auto-generated method stub
        return new Dimension(width, height);
    }
}

/**
 * @author Administrator
 * 银行对象,拥有一个阻塞队列和账户列表
 */
class MyBank {
```

```java
BlockingQueue<TransInstruction> queue;
List<Account> accounts;
int totalBalance;

public MyBank(int nAccount, int initAccountBalance) {
    try {
        Logger.getGlobal().addHandler(new FileHandler());
        //默认文件处理器,在用户主目录生成形如 javaX.log 的日志文件
    } catch (SecurityException | IOException e) {
        //TODO Auto-generated catch block
        e.printStackTrace();
    }
    accounts = new ArrayList<Account>(nAccount);              //账户列表
    totalBalance = nAccount * initAccountBalance;              //总账户金额
    queue = new ArrayBlockingQueue<TransInstruction>(30);     //初始化阻塞队列
    for (int i = 0; i < nAccount; i++) {                       //初始化账户列表
        accounts.add(new Account("Account" + i, initAccountBalance));
    }
}

/**
 * 非同步方法,没有使用锁,并非线程安全的
 */
public void trans(TransInstruction ta) {
    Account fromAccount=this.accounts.get(ta.fromAccountIndex);
    Account toAccount=this.accounts.get(ta.toAccountIndex);
    int transCash=ta.transCount;
    fromAccount.balance = fromAccount.balance - transCash;
    toAccount.balance = toAccount.balance + transCash;
    //这里使用了简单的全局日志记录

    Logger.getGlobal().info(fromAccount.name + "->"
        + toAccount.name + " =" + transCash + " total="
        + getCurrentTotalBalance());

}

public int getCurrentTotalBalance() {
    int sum = 0;
    for (Account a : accounts) {
        sum += a.balance;
    }
    return sum;
```

```java
        }
    }

/**
 * @author Administrator
 * 转账操作指令类
 */
class TransInstruction {
    int fromAccountIndex;
    int toAccountIndex;
    int transCount;
    public TransInstruction(int fromAccountIndex, int toAccountIndex, int transCount) {
        super();
        this.fromAccountIndex = fromAccountIndex;
        this.toAccountIndex = toAccountIndex;
        this.transCount = transCount;
    }

}

/**
 * @author Administrator 简单的账户类
 */
class Account {
    String name;
    int balance;

    public Account(String name, int balance) {
        this.name = name;
        this.balance = balance;
    }

    @Override
    public String toString() {
        //TODO Auto-generated method stub
        return name + "的账户余额=" + balance;
    }

}
```

3. 程序分析与注意事项

此部分包括程序解释及常见问题。

(1) 生产者线程

本例中启动 100 个线程，向阻塞队列中放置转账操作对象，线程中设置了一个死循环，让每个线程不停地生成转账操作对象，并放置到阻塞队列中。

```
for (int i = 0; i < 100; i++) {
    Runnable produce = () -> {
        while (true) {
            …
        }
    };
    Thread threadProduce = new Thread(produce);
    threadProduce.start();
}
```

(2) 消费者线程

启动 1 个线程，从阻塞队列中取转账对象，并实施转账操作。由于只有一个线程访问转账代码，所以不需要同步处理。线程中也有一个死循环，让线程不停地从阻塞对列中取得转账对象，并实施操作。由于单线程访问银行内部的账户列表，因此不需要同步机制。

```
Runnable consume = () -> {
    while (true) {
        …
    }
};
Thread threadConsume = new Thread(consume);
threadConsume.start();
```

(3) 控件更新线程

由于 Swing 不是线程安全的，因此所有对于 Swing 控件属性的设置都要放在 EventQueue.invokeLater 方法中，即使用事件分配线程更新 Swing 控件。

```
Runnable r = () -> {
    while (true) {//这个更新操作不停地进行
        //注意：任何对于 Swing 控件的操作务必放置在事件队列中！！！！
        EventQueue.invokeLater(() -> {
            value = bank.queue.size();    //获取阻塞对列中的对象数目

            statusPanel.setValue(value);
        });
        try {
            Thread.sleep(500);              //不用更新太频繁
        } catch (Exception e) {
            //TODO Auto-generated catch block
            e.printStackTrace();
```

```
            };
        Thread t = new Thread(r);
        t.start();
```

本例中使用 JProgressBar 显示队列内元素个数。

(4) 阻塞队列操作方法

阻塞队列中包含多个方法。

add：添加一个元素，如果队满了，则抛出 IllegalStateException 异常。
element：获得队头元素，如果空队，则抛出 NoSuchElementException 异常。
offer：添加一个元素，成功返回 true，如果队满返回 false。
peek：获得队头元素，如果空队返回 null。
poll：移除并返回队头元素，如果空队返回 null。
put：添加一个元素，队满则阻塞。
remove：移除并返回队头元素，如果空队，则抛出 NoSuchElementException 异常。
take：移除并返回队头元素，如果空队，则阻塞。

练习题

1. 编制程序，其功能是求：

$$M = \frac{\text{SIN}(x)}{\text{SIN}(x-y)\text{SIN}(x-z)} + \frac{\text{SIN}(y)}{\text{SIN}(y-x)(y-z)} + \frac{\text{SIN}(z)}{\text{SIN}(z-x)(z-y)} * E$$

式中，x，y，z 从键盘输入，要求用自定义异常来捕获异常，$E=5$。

2. 编制程序，其功能是：用随机函数产生 10 个 1000 以内的 7 的倍数，并从屏幕上显示其中的最大数和最小数。

3. 编制程序，其功能是：在存放一组成绩的数组中求平均成绩，将低于平均分的所有分数存放在另一数组中，并输出（要求只输出此数组中的分数）。

4. 用继承 Thread 类的方法实现一个多线程程序，该程序先后启动三个线程，每隔 2 秒打印出一条线程创建信息，然后休眠一个随机时间，最后打印出线程结束信息退出。

5. 同时显示两个时区的时钟，并将时钟的结果显示在 From 窗体中。

6. 写一个程序，该程序启动后出现一个窗口，窗口上有一个按钮，按一下这个按钮，就在窗口中随机的位置产生一个随机颜色的方块，停留 3 秒后消失。每个方块均由一个线程产生。

7. 计算 1-1/2+1/3-1/4+…+1/99-1/100 的值，并将计算结果输出到文本框中。

8. 求以下分数序列的前 10 项之和。1/1+2/3+5/8+13/21+34/55+……，其中第一项为 1/1，从第二项起其分子是前一项的分子分母之和，分母是前一项的分母和本项的分子之和，将结果显示在文本框中。

9. 键盘输入一个数，并将此数插入在一维数组中的最大数的前面。

第7章 文 件

Java 把不同来源和目标的数据都统一抽象为流,通过一系列完整的类来完成 I/O 操作。

本章将向读者介绍 Java 中 I/O 类库的特点和结构,以使读者对 I/O 类有大概的了解,然后根据访问方式的不同分别详细介绍基本文件操作、二进制文件读写、对象序列化、内存映像文件的使用等。

实践 7-1 文件基本操作

1. 实践结果

本例的运行结果如图 7-1 所示。

图 7-1 实践 7-1 的运行界面

2. 实践目的

本例介绍 Files 类的使用,包括获取文件的路径,测试文件的属性,目录操作等。

(1) 读取、写入文本文件

Files.readAllBytes:返回一个字节数组。

Files.readAllLines:返回一个 List<String>。

Files.write:可以写一个字符串到文件,也可以将行集合写到文件。

(2) 创建目录、文件

Files.createDirectory:创建新目录。

Files.createDirectories:创建各级新目录(中间目录也会被创建)。

Files.createFile:创建文件。

（3）移动、复制、删除文件

Files.copy：复制文件。

Files.move：移动文件。

Files.delete：删除文件。

Files.deleteIfExist：文件不存在不会报错，也用于删除空目录。

（4）访问目录中的项和获取文件信息

Files.readAttributes：可以获得基本文件属性的对象。

Files.list：可以获得一个Stream<Paht>对象。

3. 程序源代码

程序的源代码如下：

```java
package exp1;

import java.awt.BorderLayout;
import java.awt.Dimension;
import java.awt.EventQueue;
import java.awt.FlowLayout;
import java.awt.HeadlessException;
import java.io.File;
import java.io.IOException;
import java.nio.charset.Charset;
import java.nio.charset.StandardCharsets;
import java.nio.file.Files;
import java.nio.file.Path;
import java.nio.file.Paths;
import java.nio.file.StandardCopyOption;
import java.nio.file.StandardOpenOption;
import java.nio.file.attribute.BasicFileAttributes;
import java.util.List;
import java.util.logging.FileHandler;
import java.util.logging.Logger;
import java.util.stream.Stream;

import javax.swing.JButton;
import javax.swing.JFileChooser;
import javax.swing.JFrame;
import javax.swing.JLabel;
import javax.swing.JPanel;
import javax.swing.JScrollPane;
import javax.swing.JTextArea;
import javax.swing.plaf.FileChooserUI;
```

```java
public class Exp1 {
    public static void main(String[] args) {
        //TODO Auto-generated method stub
        EventQueue.invokeLater(() -> {
            MainFrame frame = new MainFrame("文件操作示例");
            frame.pack();
            frame.setDefaultCloseOperation(JFrame.EXIT_ON_CLOSE);
            frame.setVisible(true);
        });
    }
}

class MainFrame extends JFrame {

    private static Logger log;
    JFileChooser choose = new JFileChooser();

    public MainFrame(String arg0) throws HeadlessException {
        super(arg0);
        this.setPreferredSize(new Dimension(800, 400));
        log = Logger.getLogger("ncist.zw.java.chapter7.exp1");        //获得日志
        String userDir = System.getProperty("user.dir");              //用户工作目录
        String lineSeparator = System.getProperty("line.separator");  //系统的换行符号
        Path logFilePath = Paths.get(userDir, "exp1.log");            //获取日志文件路径
        choose.setCurrentDirectory(logFilePath.getParent().toFile()); //采用日志文件的父目录
                                                                      //作为选择器的当前目录
        try {
            FileHandler fh = new FileHandler(logFilePath.toString());
            fh.setEncoding(StandardCharsets.UTF_8.name());            //文件处理器的编码方式

            log.addHandler(fh);     //设置日志的文件处理器,日志内容会输出到文件
        } catch (SecurityException | IOException e) {
            //TODO Auto-generated catch block
            e.printStackTrace();
        }

        log.info("用户工作目录:" + userDir);
        log.info("日志位置:" + logFilePath.toString());

        //以下添加若干个按钮及其面板
        JPanel buttonPanel = new JPanel();
```

```java
buttonPanel.setPreferredSize(new Dimension(200,200));
JButton readButton = new JButton("读取日志文件"), writeButton = new JButton(
        "写入普通文本文件");

    JButton createFileButton = new JButton("创建文件"), createDirectoryButton = new JButton(
        "创建目录");
JButton copyFileButton = new JButton("复制日志文件"), moveFileButton = new JButton(
        "移动日志文件副本"), deleteFileButton = new JButton("删除指定文件");
JButton readLogFileInfo = new JButton("读取指定目录中的文件信息");

buttonPanel.add(readButton);
buttonPanel.add(writeButton);

buttonPanel.add(createFileButton);
buttonPanel.add(createDirectoryButton);
buttonPanel.add(copyFileButton);
buttonPanel.add(moveFileButton);
buttonPanel.add(deleteFileButton);
buttonPanel.add(readLogFileInfo);

//以下添加一个TextArea及其父面板
JPanel contentPanel = new JPanel();
contentPanel.setLayout(new BorderLayout());

JTextArea textArea = new JTextArea();
JScrollPane scrollP = new JScrollPane(textArea);
contentPanel.add(scrollP, BorderLayout.CENTER);
textArea.append("Hello!");

/*
 * 以下设置按钮的监听器
 */

readButton.addActionListener((event) -> {
        //按字节读取日志文件
        try {
            byte[] bytes = Files.readAllBytes(logFilePath);            //读成字节数组
            String str = new String(bytes, StandardCharsets.UTF_8);    //注意编码方式
            textArea.setText(str);
        } catch (Exception e) {
            //TODO Auto-generated catch block
            e.printStackTrace();
```

```java
        }
        textArea.append("--------------------------------\n");

        //按行读取日志文件
        try {
            List<String> lines = Files.readAllLines(logFilePath);
            for (String line : lines) {
                textArea.append(line + lineSeparator);
            }

        } catch (Exception e) {
            //TODO Auto-generated catch block
            e.printStackTrace();
        }

    });

writeButton.addActionListener((event) -> {
    choose.setFileSelectionMode(JFileChooser.FILES_ONLY);     //只能选择文件
    if (JFileChooser.APPROVE_OPTION == choose.showDialog(this,
            "写入文件"))  {                                    //用户选择了文件
        File f = choose.getSelectedFile();
        log.info("用户要求写入:" + f.toString());
        try {
            Files.write(
                    f.toPath(),
                    textArea.getText().getBytes(
                            StandardCharsets.UTF_8),
                    StandardOpenOption.CREATE);
            //这里采用将文本区的文字转化成特定编码方式的字节,之后写入
            //文件
        } catch (Exception e) {
            //TODO Auto-generated catch block
            e.printStackTrace();
        }
    }
});

createFileButton.addActionListener((event) -> {
    choose.setFileSelectionMode(JFileChooser.FILES_ONLY);     //只能选择文件
    if (JFileChooser.APPROVE_OPTION == choose
            .showDialog(this, "创建"))  {                     //用户选择了文件
```

```java
            File f = choose.getSelectedFile();
            log.info("用户要求创建:" + f.toString());
            try {
                Files.createFile(f.toPath());        //如果文件存在则抛出异常
                log.info(f.toString() + "是空文件!");
            } catch (Exception e) {
                //TODO Auto-generated catch block
                log.severe(f.toString() + "文件已存在!");    //捕获异常后记录异常
            }
        }
    });
    createDirectoryButton.addActionListener((event) -> {
        choose.setFileSelectionMode(JFileChooser.FILES_ONLY);    //只能选择文件
        if (JFileChooser.APPROVE_OPTION == choose.showDialog(this,
                "创建目录")) {                                    //用户选择了文件
            File f = choose.getSelectedFile();
            log.info("用户要求创建目录:" + f.toString());
            try {
                Files.createDirectory(f.toPath());
                //如果希望创建中间各级目录请使用createDirectories
                log.info(f.toString() + "目录创建完成!");
            } catch (Exception e) {
                //TODO Auto-generated catch block
                log.severe(f.toString() + "目录创建失败!");    //捕获异常后记录异常
            }
        }
    });
    copyFileButton.addActionListener((event) -> {
        //复制文件
        Path logCopy = Paths.get(userDir, "logCopy.log");
        try {
            Files.copy(logFilePath, logCopy,
                    StandardCopyOption.REPLACE_EXISTING,
                    StandardCopyOption.COPY_ATTRIBUTES);    //复制日志文件,
                                                            //文件名固定
            //复制时目标存在则替换,并复制文件属性
            log.info("复制了日志文件!");
        } catch (Exception e) {
            //TODO Auto-generated catch block
            log.info("复制日志文件,失败!");
        }
    });
```

```java
moveFileButton.addActionListener((event) -> {
    choose.setFileSelectionMode(JFileChooser.DIRECTORIES_ONLY);    //只能选择目录
    Path logCopyPath = Paths.get(userDir, "logCopy.log");
    if (JFileChooser.APPROVE_OPTION == choose.showDialog(this,
        "移动到")) {                                                //用户选择了文件
        File f = choose.getSelectedFile();

        log.info("用户要求移动日志副本到目录:" + f.toString());
        Path targetPath = Paths.get(f.toString(), "logCopy.log");
        try {

            Files.move(logCopyPath, targetPath,
                StandardCopyOption.ATOMIC_MOVE);
            //移动日志副本,设置过程的原子性

            log.info("移动文件:" + logCopyPath + " 到 " + targetPath);
        } catch (Exception e) {
            //TODO Auto-generated catch block
            log.severe("移动文件:" + logCopyPath + " 到 " + targetPath
                + " =》失败!");                                     //捕获异常后记录异常
            log.severe(e.toString());
        }
    }
});
deleteFileButton.addActionListener((event) -> {
    choose.setFileSelectionMode(JFileChooser.FILES_ONLY);          //只能选择文件
    if (JFileChooser.APPROVE_OPTION == choose
        .showDialog(this, "删除")) {                                //用户选择了文件
        File f = choose.getSelectedFile();
        log.info("用户要求删除文件:" + f.toString());
        try {
            Files.delete(f.toPath());          //如果文件不存在则抛出异常
            //如果希望不出异常,可以用 Files.deleteIfExists
            log.info(f.toString() + "文件删除!");
        } catch (Exception e) {
            //TODO Auto-generated catch block
            log.severe(f.toString() + "文件删除失败!");//捕获异常后记录异常
        }
    }
});
readLogFileInfo.addActionListener((event) -> {
    choose.setFileSelectionMode(JFileChooser.DIRECTORIES_ONLY);    //只能选择目录
    if (JFileChooser.APPROVE_OPTION == choose.showDialog(this,
```

```java
                        "选定该目录"))){                    //用户选择了文件
            textArea.setText("");
            File f = choose.getSelectedFile();
            log.info("用户要求访问目录项:" + f.toString());
            try(Stream<Path> items = Files.list(f.toPath())){  //获得流对象
                items.forEach((path) -> {                    //流处理
                    BasicFileAttributes atts;                //文件的基本属性集
                    try{
                        atts = Files.readAttributes(path,
                                BasicFileAttributes.class);
                        textArea.append(path + "字节大小:" + atts.size()+"\n");
                        textArea.append(path + "创建时间:" + atts.creationTime()+"\n");
                        textArea.append(path + "是否为目录:" + atts.isDirectory()+"\n");
                        textArea.append(path + "是否为常规文件:" + atts.isRegularFile()+"\n");
                        textArea.append(path + "是否是符号链接:" + atts.isSymbolicLink()+"\n");
                        textArea.append(path + "最近访问时间:" + atts.lastAccessTime()+"\n");
                        textArea.append(path + "最接近修改时间:" + atts.lastModifiedTime()+"\n");
                    }catch(Exception e){
                        //TODO Auto-generated catch block
                        e.printStackTrace();
                    }
                });
            }catch(IOException e){
                e.printStackTrace();
            }
        }
    });
    add(buttonPanel, BorderLayout.EAST);
    add(contentPanel, BorderLayout.CENTER);
    }
}
```

4. 程序分析与注意事项

此部分包括程序解释及常见问题。

（1）系统属性

System.getProperty("user.dir")获得用户工作目录。

System.getProperty("line.separator")获得系统的换行符号。

(2) 日志

Logger. getLogger("ncist. zw. java. chapter7. exp1")创建日志。

log. addHandler(fh)给日志制定一个文件处理器,从而可以将日志内容写入文件。

(3) 字节数组与字符串的相互转换

byte[ ] bytes = Files. readAllBytes(logFilePath)文件读成字节数组。

String str = new String(bytes, StandardCharsets. UTF_8)从字节数组构造字符串,指定编码方式。

(4) 文件复制和移动时操作选项

Files. move(., ., StandardCopyOption. ATOMIC_MOVE)移动过程的原子性。

Files. copy(.., ., StandardCopyOption. REPLACE_EXISTING, StandardCopyOption. COPY_ATTRIBUTES)覆盖已存在目标文件,并保留源文件的文件属性。

实践 7-2 二进制文件和对象序列化

1. 实践结果

本例的运行结果如图 7-2 所示。

```
原始的tom=>Customer [name=tom, b=116, age=18, salary=2900.98, c=T, id=123456,
原始的tom写入二进制文件tom.dat
从tom.dat读取的recoveryTom=>Customer [name=tom, b=116, age=18, salary=2900.98
利用RandomAccessFile类读tom.dat,从中读取LoginInfo=>LoginInfo [loginName=tom12:
原始的tom写入对象文件tom.object
利用ObjectInputSteam,从tom.object中读取对象=>Customer [name=tom, b=116, age=18,
tom , tom2 不是同一个对象!
tom.loginInfo , tom2.loginInfo 不是同一个对象!
```

图 7-2 实践 7-2 的控制台输出

2. 实践目的

本例介绍 DataInput 接口、DataOutput 接口、RandomAccessFile 类、ObjectInputStream 类和 ObjectOutputStream 类的使用方法。

DataInput 接口定义了二进制格式读取数组、字符、整数等方法,主要方法有:readChars、readByte、readInt、readLong、readShort、readFloat、readBoolean 等。DataInputStream 类实现了 DataInput 接口。

DataOutput 接口定义了二进制格式写数组、字符、整数等方法,对于每种给定的类型,所占的空间都是固定的。主要方法有:writeChars、writeByte、writeInt、writeLong、writeShort、writeFloat、writeBoolean 等。DataOutputStream 类实现了 DataOutput 接口。

RandomAccessFile 类,同时实现了 DataInput 接口和 DataOutput 接口,可以在文件中的任意位置读取和写入数据。其 seek 方法可以将文件指针设置到文件中的任意字节,它的参数是一个 long 类型的整数。

ObjectInputStream 类和 ObjectOutputStream 类,可以将任何对象写出到输出流中,并且在之后读回这个对象。对象序列化的要求很简单,只需要类实现 Serializable 接口,但却不需要实现任何接口方法。ObjectInputSteam 实现了 ObjectInput 接口,而 ObjectInput 接口继承自

DataInput；ObjectOutputStream 实现了 ObjectOutput 接口，ObjectOutput 接口是 DataOutput 的子接口。

3. 程序源代码

程序的源代码如下：

```java
package exp2;

import java.io.DataInput;
import java.io.DataInputStream;
import java.io.DataOutput;
import java.io.DataOutputStream;
import java.io.FileInputStream;
import java.io.FileNotFoundException;
import java.io.FileOutputStream;
import java.io.IOException;
import java.io.ObjectInputStream;
import java.io.ObjectOutputStream;
import java.io.RandomAccessFile;
import java.io.Serializable;

public class Exp2 {
    public static void main(String[] args) {
        //TODO Auto-generated method stub
        LoginInfo lg=new LoginInfo("tom123","456");      //登录信息对象
        Customer tom=new Customer("tom",(byte)'t',18,2900.98,'T',123456,lg);//客户对象
        System.out.println("原始的 tom =>"+tom.toString());
        //tom 对 lg 形成一种引用关系
        Customer recoveryTom=new Customer();    //一个从数据文件恢复出来的 Customer 对象

        //将 tom 进行二进制输出,利用 DataOutput 接口的方法
        try(DataOutputStream out=new DataOutputStream(new FileOutputStream("tom.dat"))){
            tom.toDataFile(out);
            System.out.println("原始的 tom 写入二进制文件 tom.dat");
        }
        catch (IOException e1) {
            //TODO Auto-generated catch block
            e1.printStackTrace();
        }

        //利用 DataInput 接口方法读取二进制文件 tom.dat,从中读取信息,恢复为 recoveryTom
        //对象
        try(DataInputStream in=new DataInputStream(new FileInputStream("tom.dat"))){
```

```
                recoveryTom.fromDataFile(in);
                System.out.println("从 tom.dat 读取的 recoveryTom=>"+recoveryTom);
            } catch (IOException e) {
                //TODO Auto-generated catch block
                e.printStackTrace();
            }
            //利用随机文件读取类 RandomAccessFile,从 tom.dat 中读取 logininfo 对象
            //RandomAccessFile 类同时实现了 DataInpu 和 DataOutput 接口
            try(RandomAccessFile in=new RandomAccessFile("tom.dat","r")){

                LoginInfo tempLg=Customer.readLoginInfo(in);
                System.out.println("利用 RandomAccessFile 类读取 tom.dat,从中读取 LoginInfo=>"+tempLg);
            } catch (IOException e) {
                //TODO Auto-generated catch block
                e.printStackTrace();
            }

            //利用 ObjectOutputSteam 将对象 tom 输出为文件
            try(ObjectOutputStream objectOut=new ObjectOutputStream(new FileOutputStream("tom.object"))){
                objectOut.writeObject(tom);//注意 writeObject 只能保存对象,不能用于基本数据类
                                            //型(仍然要使用 writeInt 等类似方法)
                System.out.println("原始的 tom 写入对象文件 tom.object");
            } catch (IOException e) {
                //TODO Auto-generated catch block
                e.printStackTrace();
            }
            Customer tom2=null;
            //利用 ObjectInputStream 读入文件 tom.object,并存储为 tom2
            try(ObjectInputStream objectIn=new ObjectInputStream(new FileInputStream("tom.object"))){
                tom2=(Customer)objectIn.readObject();
                System.out.println("利用 ObjectInputSteam,从 tom.object 中读取对象=>"+tom2);
            } catch (IOException e) {
                //TODO Auto-generated catch block
                e.printStackTrace();
            } catch (ClassNotFoundException e) {
                //TODO Auto-generated catch block
                e.printStackTrace();
            }
            //注意 tom2 与 tom 虽然所有的属性值全都一样,然而属于不同的对象
            if(tom==tom2){
```

```java
                System.out.println("tom, tom2 是同一个对象!");
            } else {
                System.out.println("tom, tom2 不是同一个对象!");
            }
            //tom, tom2 的 loginInfo 指向的也不是同一个对象
            if(tom.loginInfo == tom2.loginInfo) {
                System.out.println("tom.loginInfo, tom2.loginInfo 是同一个对象!");
            } else {
                System.out.println("tom.loginInfo, tom2.loginInfo 不是同一个对象!");
            }
    }
}
/**
 * @author Administrator
 * 普通的客户类,
 * Serializable 接口用于进行对象的输入输出序列化,程序员不需要实现任何方法
 */
class Customer implements Serializable {
    /**
     * serialVersionUID 是类的版本号
     */
    private static final long serialVersionUID = 1L;
    final static int NAME_LENGTH = 20;
    String name;
    byte b;
    int age;
    double salary;
    char c;
    long id;
    LoginInfo loginInfo;
    public Customer(String name, byte b, int age, double salary, char c,
            long id, LoginInfo loginInfo) {
        super();
        this.name = name;
        this.b = b;
        this.age = age;
        this.salary = salary;
        this.c = c;
        this.id = id;
        this.loginInfo = loginInfo;
    }
    public Customer() {
```

```
        //TODO Auto-generated constructor stub
    }


    @Override
    public String toString() {
        return "Customer [name=" + name + ", b=" + b + ", age=" + age
            + ", salary=" + salary + ", c=" + c + ", id=" + id
            + ", loginInfo=" + loginInfo + "]";
    }
    /**
     * 将字符出串以字符为单位写入 DataOutput 流, 保存为二进制。
     * @param s 拟输出的字符串
     * @param fixedLength 输出的固定长度, 当 s.length<fixedLength 时, 补充 0
     * @param out 输出流
     * @throws IOException
     */
    private static void toFixedLengthString(String s, int fixedLength, DataOutput out) throws IOException{
        for(int i=0;i<fixedLength;i++){
            if(i<s.length()){
                out.writeChar(s.charAt(i));
            }else{
                out.writeChar(0);              //输出 0 值, 而不是输出'0'
            }
        }
    }
    /** 从二进制文件中利用 DataInput 以字符为单位读取固定长度的字符串, 舍弃补充的 0
     * @param fixedLength 需要读取的固定长度
     * @param in 输入流
     * @return 获得的字符串
     * @throws IOException
     */
    private static String fromFixedLengthChar(int fixedLength,DataInput in) throws IOException{
        StringBuilder str=new StringBuilder(fixedLength);//利用 StringBuilder, 并设置初始容量
        char c;
        for(int i=0;i<fixedLength;i++){

            if((c=in.readChar())!=0){
                str.append(c);
            }

        }
        return str.toString();
```

}

/**
 * 将 Customer 对象输出到 DataOutput, 以二进制形式
 * @param out 输出流
 * @throws IOException
 */
public void toDataFile(DataOutput out) throws IOException{
    toFixedLengthString(name, NAME_LENGTH, out);
    out.writeByte(b);
    out.writeInt(age);
    out.writeDouble(salary);
    out.writeChar(c);
    out.writeLong(id);
    toFixedLengthString(loginInfo.loginName, LoginInfo.LOGIN_NAME_LENGTH, out);
    toFixedLengthString(loginInfo.password, LoginInfo.PASSWORD_LENGTH, out);
}

/** 从输入流 DataInput 中以二进制形式形式读入 Customer 对象
 * @param in 输入流
 * @throws IOException
 */
public void fromDataFile(DataInput in) throws IOException{
    this.name = fromFixedLengthChar(NAME_LENGTH, in);
    this.b = in.readByte();
    this.age = in.readInt();
    this.salary = in.readDouble();
    this.c = in.readChar();
    this.id = in.readLong();
    String loginName = fromFixedLengthChar(LoginInfo.LOGIN_NAME_LENGTH, in);
    String pass = fromFixedLengthChar(LoginInfo.PASSWORD_LENGTH, in);
    this.loginInfo = new LoginInfo(loginName, pass);
}

/** 从 Customer 的二进制文件中, 仅读取 LoginInfo 对象。跳过了前面的所有其他属性
 * 利用 RandomAccessFile 的随机访问特性, 跳过文件中的指定字节数
 * @param file 随机访问文件
 * @return
 * @throws IOException
 */
public static LoginInfo readLoginInfo(RandomAccessFile file) throws IOException{
    file.seek(NAME_LENGTH*(2)+1+4+8+2+8);//跳过字节
    //读取 LoginInfo 的两个属性值
```
```

```java
            String loginName = fromFixedLengthChar(LoginInfo.LOGIN_NAME_LENGTH, file);
            String pass = fromFixedLengthChar(LoginInfo.PASSWORD_LENGTH, file);
            LoginInfo lg = new LoginInfo(loginName, pass);
            return lg;
        }
    }
    /**
     * @author Administrator
     * 登录信息类，也支持对象序列化
     */
    class LoginInfo implements Serializable{
        /**
         * serialVersionUID 是类的版本号
         */
        private static final long serialVersionUID = 1L;
        final static int LOGIN_NAME_LENGTH = 20;
        final static int PASSWORD_LENGTH = 30;
        public LoginInfo(String loginName, String password) {
            super();
            this.loginName = loginName;
            this.password = password;
        }
        String loginName;
        String password;
        @Override
        public String toString() {
            return "LoginInfo [loginName=" + loginName + ", password=" + password
                    + "]";
        }
    }
}
```

4. 程序分析与注意事项

利用 DataInput、DataOutput 接口进行二进制数据读写时，应当注意数据的读顺序和写顺序的一致，否则会读取到错误的信息。

RandomAccessFile 类可以对文件进行随机的读写，但在利用 seek 方法设定文件指针位置时，要精确计算好字节位置。例如 file.seek(NAME_LENGTH*(2)+1+4+8+2+8)，表明跳过姓名字符串的字节长度（1 个 char 为 2 字节），以及后续的各个属性数据类型的字节长度之和（int 为 4 字节，long 和 double 为 8 字节）。

ObjectOutputStream 和 ObjectInputStream 支持将对象输出到流或从流中读取对象，对于基本数据类型则仍然要采用 DataInput 和 DataOutput 接口中的方法。

对象序列化时，每个对象都关联一个序列号，因此同一个对象不会被保存两次。基于这种特性，当对象间存在关联关系时，对象网络在序列化时各个对象都能够被方便地存储和加载。由于类存在演化和版本的情况，因此不同版本的同一个类可以指定一个值相同的 serialVerisonUID 静态数据成员，这样序列化系统就可以读入这个类对象的不同版本。

实践 7-3　内存映射文件

1. 实践结果

本例正常运行时的结果如图 7-3 所示。本例主要是利用普通输入流、带缓冲的输入流和内存映射对大文件的字节数进行统计，从而比较它们的执行速度。

```
内存映射.....
内存映射花费 17毫秒,读取字节=18173050
BufferedInputStream.....
BufferedInputStream花费 301毫秒,读取字节=18173050
InputStream.....
InputStream花费 24235毫秒,读取字节=18173050
```

图 7-3　实践 7-3 的控制台输出

2. 实践目的

现代操作系统中可以利用虚拟内存将文件或文件的某个部分映射到内存中，之后可以像数组一样对它进行修改或访问，这比常见的缓冲文件方式更加高效。

① FileChannel 类可以获得磁盘文件的通道，这是对于文件的高级抽象，帮助程序员访问内存映射、文件数据传递和文件锁等系统特性。例如：

FileChannel channel = (FileChannel.open(file))

② FileChannel 类的 map 方法可以获得与通道对应的 ByteBuffer 对象，利用它可以读写数据。例如：

MappedByteBuffer byteBuffer = channel.map(MapMode.READ_ONLY, 0, size);

其中的参数用于设置映射区域和映射模式。

3. 程序源代码

程序的源代码如下：

```
package exp3;

import java.io.BufferedInputStream;
import java.io.IOException;
import java.io.InputStream;
import java.nio.MappedByteBuffer;
import java.nio.channels.FileChannel;
import java.nio.channels.FileChannel.MapMode;
import java.nio.file.Files;
import java.nio.file.Path;
import java.nio.file.Paths;
```

```java
public class Exp3 {
    /**
     * 统计文件字节数
     * @param file
     * @return
     * @throws IOException
     */
    public static long getSumWithBufferedInputStream( Path file)
            throws IOException {
        long sum = 0;
        try (InputStream in = new BufferedInputStream(      //带缓冲区的输入流
                Files.newInputStream(file))) {
            int c;
            while ((c = in.read()) != -1) {
                sum ++;
            }
        }
        return sum;
    }

    /**
     * 统计文件字节数
     * @param file
     * @return
     * @throws IOException
     */
    public static long getSumWithInputStream( Path file) throws IOException {
        long sum = 0;
        try (InputStream in = (Files.newInputStream(file))) {    //非缓冲
            int c;
            while ((c = in.read()) != -1) {
                sum ++;
            }
        }
        return sum;
    }

    /**
```

```java
 * 统计文件字节数
 * @param file
 * @return
 * @throws IOException
 */
public static long getSumWithFileChannel(Path file) throws IOException {
    long sum = 0;
    int c = 0;
    try (FileChannel channel = (FileChannel.open(file))) {//
        int size = (int) channel.size();                              //与通道关联的文件大小
        MappedByteBuffer byteBuffer = channel.map(MapMode.READ_ONLY, 0,
                size);                                                //设置映射区域和映射模式
        for (int i = 0; i < size; i++) {
            c = (int) byteBuffer.get(i);
            sum++;
        }

    }
    return sum;
}

public static void main(String[] args) {
    //TODO Auto-generated method stub
    System.out.println("内存映射.....");
    long start = System.currentTimeMillis();
    long sum = 0;
    try {
        sum = getSumWithFileChannel(Paths.get("tools.jar"));
    } catch (IOException e) {
        //TODO Auto-generated catch block
        e.printStackTrace();
    }
    long end = System.currentTimeMillis();
    System.out.println("内存映射花费 " + (end - start) + "毫秒,读取字节=" + sum);
    ////
    System.out.println("BufferedInputStream.....");
    start = System.currentTimeMillis();
    sum = 0;
    try {
        sum = getSumWithBufferedInputStream(Paths.get("tools.jar"));
    } catch (IOException e) {
        //TODO Auto-generated catch block
        e.printStackTrace();
```

```
        }
        end = System.currentTimeMillis();
        System.out.println("BufferedInputStream 花费 " + (end - start) + "毫秒,读取字节="
                + sum);
        ///
        System.out.println("InputStream....");
        start = System.currentTimeMillis();
        sum = 0;
        try {
            sum = getSumWithInputStream(Paths.get("tools.jar"));
        } catch (IOException e) {
            //TODO Auto-generated catch block
            e.printStackTrace();
        }
        end = System.currentTimeMillis();
        System.out.println("InputStream 花费 " + (end - start) + "毫秒,读取字节=" + sum);
    }
}
```

4. 程序分析与注意事项

本程序利用不同的三种方法对一个较大的文件（jre 中的 tools.jar）进行读取，记录总字节数，以此考察不同方法的执行效率。

时间记录方法：在每一种方法执行前后，利用 System.currentTimeMillis() 获得当前时间，然后利用 (end - start) 计算执行时间差。

对于 ByteBuffer，可以采用顺序遍历，也可以使用随机访问，例如：

```
while(buffer.hasRemaining()){
    byte b=buffer.get();
}
for(int i=0;i<buffer.limit();i++){
    byte b=buffer.get(i);
}
```

练习题

1. 编制程序实现：计算"2*4*6+8*10*12+…+26*28*30"的值 s，最后把结果输出到"C:\save.txt"文件中。
2. 编制程序实现：某经济发达城市的工业产值每年增长 8%，编程计算多少年（变量为 N）后产值可以翻两番（基数的 4 倍），并把结果 N 存入文件"C:\save.txt"中。
3. 从"D:\hao.tex"中读取数据，将结果显示在文本框中。
4. 将文件"zhou.txt"传送到"E:\hao\de.txe"中。
5. 编制程序实现：统计出 300 以内既不能被 5 整除，也不能被 7 整除的数的个数 y，最后把结果 y 输出到文件 "save.txt"中。

6. 编制程序实现：打开一个随机文件 "save.txt"，存放读入的 5 个学生的学号、姓名和成绩，最后屏幕输出成绩最高和最低的学生的信息。

7. 编制程序实现：打开一个顺序文件 "file.txt"，读入一组单个的字符。如是英文字母，将大（小）写变换为小（大）写，其他字符不变，然后逐个显示成一个句子。

8. 编制程序实现：从文件 "File.txt" 读入的若干个字符中，统计数字、大写字母、小写字母、非字母字符的个数，并将统计结果分别存入文件 "Save.txt" 中。

9. 编制程序实现：输出 1000 以内 3 的奇数倍数，要求每行输出 5 个数，并求它们的和，最后存入名为 "save.txt" 的顺序文件中。

10. 编制程序实现：在屏幕上显示一幅动画效果，并循环播放背景音乐。

11. 编制程序实现：读入一个顺序文件，该文件每条记录的三个数据项分别是职工姓名、基本工资和奖金；在读入每条记录后，将每个职工的基本工资增加百分之十后得到新的数据记录。

12. 编制程序实现：在磁盘上创建一个电话号码文件，存放单位名称与单位电话号码。程序反复地从键盘输入单位名称与电话号码，并写到磁盘文件 "TEL" 中，直到输入的单位名称是 "DONE" 为止。

第 8 章 网络编程基础

本章将向读者讲述网络编程基础的知识，并利用 Socket 创建支持多线程的服务器程序和客户端程序；利用 URL 类和 URLConnection 访问 Web 服务器上的资源；利用 JavaMail 发送邮件。

实践 8-1 Socket 与 ServerSocket 编程

1. 实践结果

本例的运行结果如图 8-1、图 8-2 和图 8-3 所示。

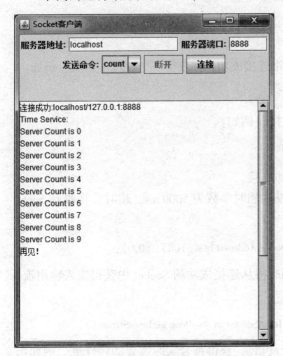 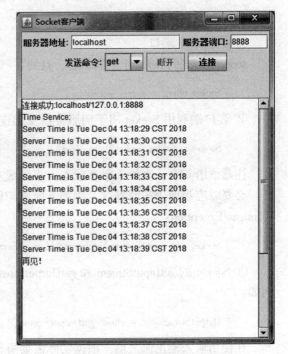

图 8-1 客户端 1 的运行界面　　　　图 8-2 客户端 2 的运行界面

本例使用 ServerSocket 构建了一个时间服务程序，用于向客户端提供服务器的当前时间，服务器利用了线程池和多线程，支持多个客户端的连接。同时对于每一个客户端发送的指令数进行统计，利用一个线程专门负责显示每一个客户端的指令数。

利用 Socket 构建了一个客户端，用于向时间服务器发送指令，并接收消息。指令分为 get 和 count。get 用于查询时间，count 用于查询当前通信发送了多少指令。

2. 实践目的

本例介绍怎样构建一个支持多客户端的服务器程序，以及构建一个客户端程序。

① 服务器端利用 ServerSocket 用于在计算机的指定端口上接收客户机的连接，例如：

```
收到连接..
-->get        ?-?-?-
?-            -->count
收到连接..     -->get
-->count      输出结束。。。
-->get        9-?-?-
?-?-          -->count
收到连接..     -->get
-->count      9-?-?-
-->get        输出结束。。。
-->get        -->get
?-?-?-        9-10-?-
-->count      -->get
-->get        9-10-?-
-->get        输出结束。。。
?-?-?-        9-10-11-
```

图 8-3 服务器端（支持多客户端）的控制台输出

```
server = new ServerSocket(8888);
```

将监听 8888 端口。

accept 方法将持续处于阻塞状态，直到客户端连接成功，并获得一个 Socket 对象，例如：

```
Socket client = server.accept();
```

② 客户端利用 Socket 用于向服务器发起连接，例如：

```
Socket s = new Socket(host, port);
```

注意该语句会一直处于阻塞状态，直到成功连接。

必要时应当使用超时参数。例如下面语句设置超时参数为 5000 ms，超时后将收到 SocketTimeoutException：

```
Socket s = new Socket();s.connect(new InetSocketAddress(host, port), 5000);
```

③ Socket 的 getInputStream 和 getOutputStream 将从连接成功的 Socket 中获得输入输出流。例如：

```
OutputStream out = client.getOutputStream();InputStream in = client.getInputStream()
```

在使用输入输出流之前，根据实际需要进行包装，获得符合实际需要的特殊流。例如：

```
Scanner input = new Scanner(in, "UTF-8");PrintWriter output = new PrintWriter(new OutputStreamWriter(out,"UTF-8"), true);
```

套接字在使用完后，需要使用 close 方法进行关闭。因此通常利用带资源的 try 语句块。

④ 线程池可以为程序提供多线程管理的支持，本例使用了固定大小的线程池，100 个线程。

构建线程池：

```
ExecutorService threadPool = Executors.newFixedThreadPool(100);
```

启动线程：

Future<Integer> result = threadPool. submit(tService);

⑤ Callable 接口和 Runnable 接口类似，除了可以构建线程外，还可以获得线程的执行结果，即具有返回值。本例中 class TimeService implements Callable<Integer>，作为时间服务实现了 Callable 接口，并要求线程执行完成后返回一个 Integer 的结果。Callable 接口中包含 call 方法，方法体内放置线程执行代码，该方法具有返回值。例如 public Integer call()。

⑥ Future 接口，用于保存异步计算结果，在线程执行后，可以利用 Future 接口的对象获得执行结果。

get 方法用于获得线程计算结果，如果没有计算完成，则一直处于阻塞状态。

isDone 方法用于查询计算是否完成。

⑦ 更新 Swing 控件时，务必利用事件队列，例如：

EventQueue. invokeLater(() -> {textArea. append(str + "\n");});

3. 程序源代码

服务器端程序的源代码如下：

```
package exp1;

import java. io. IOException;
import java. io. InputStream;
import java. io. OutputStream;
import java. io. OutputStreamWriter;
import java. io. PrintWriter;
import java. net. ServerSocket;
import java. net. Socket;
import java. time. LocalTime;
import java. util. ArrayList;
import java. util. Calendar;
import java. util. Date;
import java. util. List;
import java. util. Locale;
import java. util. Scanner;
import java. util. concurrent. Callable;
import java. util. concurrent. ExecutionException;
import java. util. concurrent. ExecutorService;
import java. util. concurrent. Executors;
import java. util. concurrent. Future;

public class MyServer {

    public static void main(String[] args) {
        ServerSocket server;
        //服务器端采用多线程，因此采用线程池。防止并发过高时创建过多线程耗尽资源
```

```java
ExecutorService threadPool = Executors.newFixedThreadPool(100);
//固定大小的线程池，空闲程也会被保留

List<Future<Integer>> serverThreadResult = new ArrayList<Future<Integer>>();
//保留每个线程的执行结果

try {

    //接下来启动一个线程，专门在错误输出(仅为了方便，而非错误)中打印时间服务
    //TimeService 中的执行结果，也就是接收了多少指令
    Runnable r = () -> {
        while (true) {
            try {
                Thread.sleep(1000);//每隔 1 秒
                String str="";
                for (Future<Integer> f : serverThreadResult) {
                    if (f.isDone())//Future 的 isDone 方法可以查询是否线程结束
                        str+=(f.get() + "-");//get 方法会被阻塞，所以这里判断一下
                    else
                        str+=("?" + "-");
                }
                System.err.print(str+"\n");
            } catch (Exception e) {
                //TODO Auto-generated catch block
                e.printStackTrace();
            }//错误输出中打印每个服务线程执行的命令数
        }
    };
    Thread t = new Thread(r);
    t.start();//启动结果查询线程

    server = new ServerSocket(8888);//服务器套接字，初始化为监听 8888 端口
    System.out.println("服务器启动...");
    while (true) {//不停循环，直到服务器关闭
        Socket client = server.accept();//阻塞方法，等待客户连接
        System.out.println("收到连接..");
        TimeService tService = new TimeService(client);//创建时间服务对象
        Future<Integer> result = threadPool.submit(tService);//放入线程池
        serverThreadResult.add(result);//将结果存入线程执行结果列表
    }
} catch (Exception e) {
    //TODO Auto-generated catch block
    e.printStackTrace();
```

```java
        }

    }
}

/**
 * @author Administrator 时间服务,实现了 Callable 接口,用法与 Runnable 类似,但可以在线程
 * 结束后,返回一个结果
 */
class TimeService implements Callable<Integer> {
    private Socket client;

    public TimeService(Socket client) {
        super();
        this.client = client;
    }

    @Override
    public Integer call() throws Exception {
        int count = 0;
        try (OutputStream out = client.getOutputStream();
             InputStream in = client.getInputStream()) {
            //从 Socket 中获得输入输出流,并适当包装一下
            Scanner input = new Scanner(in, "UTF-8");
            PrintWriter output = new PrintWriter(new OutputStreamWriter(out,
                    "UTF-8"), true);//自动刷新缓冲区
            output.println("Time Service:");

            while (input.hasNextLine()) {//hasNextLine 会阻塞,等待每一行输入
                String s = input.nextLine();
                System.out.println("-->" + s);
                if (s.equals("count")) {//解析指令,并向客户端写入数据
                    output.println("Server Count is " + count);
                } else if (s.equals("get")) {
                    output.println("Server Time is " + new Date());
                } else {
                    output.println("Time Service:");
                }
                //休息 1 秒
                Thread.sleep(1000);
                count++;
            }
            //客户端的 Socket 调用 shutdownOutput()后,服务器端知道数据传输结束了
```

```
        //服务器端写回信息
        output.println("再见!");
        output.flush();
        System.out.println("输出结束...");

    } catch (IOException | InterruptedException e) {
        //TODO Auto-generated catch block
        e.printStackTrace();
    }
    //线程结束了，返回当前线程全部接收到的指令
    return Integer.valueOf(count);
  }

}
```

客户端程序代码如下：

```
package exp1;

import java.awt.BorderLayout;
import java.awt.Dimension;
import java.awt.EventQueue;
import java.awt.HeadlessException;
import java.awt.TextArea;
import java.io.IOException;
import java.io.OutputStreamWriter;
import java.io.PrintWriter;
import java.net.InetAddress;
import java.net.InetSocketAddress;
import java.net.Socket;
import java.net.SocketAddress;
import java.net.SocketTimeoutException;
import java.util.Scanner;

import javax.swing.JButton;
import javax.swing.JComboBox;
import javax.swing.JFrame;
import javax.swing.JLabel;
import javax.swing.JPanel;
import javax.swing.JScrollPane;
import javax.swing.JTextArea;
import javax.swing.JTextField;

public class ExpOne {
```

```java
    public static void main(String[] args) {
        //TODO Auto-generated method stub
        EventQueue.invokeLater(() -> {
            MyClient client = new MyClient();
            client.setDefaultCloseOperation(JFrame.EXIT_ON_CLOSE);
            client.pack();
            client.setVisible(true);
        });
    }

}

class MyClient extends JFrame {
    JTextArea textArea = new JTextArea(30, 30);//为了在内部类里使用,所以放在这里
    JComboBox<String> sendComm = new JComboBox<String>();
    Thread tRe = null;

    public MyClient() {
        super("Socket 客户端");
        //TODO Auto-generated constructor stub
        this.setPreferredSize(new Dimension(400, 500));
        JPanel bPanel = new JPanel();
        bPanel.setPreferredSize(new Dimension(400, 100));
        bPanel.add(new JLabel("服务器地址:"));
        JTextField serverAddress = (JTextField) bPanel.add(new JTextField(
                "localhost", 15));//服务器地址
        bPanel.add(new JLabel("服务器端口:"));
        JTextField serverPort = new JTextField("8888", 5);//端口
        bPanel.add(serverPort);

        bPanel.add(new JLabel("发送命令:"));

        sendComm.addItem("get");//客户端发送指令的集合
        sendComm.addItem("count");
        sendComm.setSelectedIndex(0);
        bPanel.add(sendComm);

        JButton connectNormalButton = new JButton("连接");

        JButton disconnectButton = new JButton("断开");
        disconnectButton.setEnabled(false);//"断开"按钮默认不可用
        bPanel.add(disconnectButton);

        this.add(new JScrollPane(textArea), BorderLayout.CENTER);
```

```java
        bPanel.add(connectNormalButton);
        //添加按钮监听器
        connectNormalButton.addActionListener((event) -> {
            String address = serverAddress.getText();

            int port = Integer.parseInt(serverPort.getText().trim());
            //创建一个接收线程,并利用Socket与服务器通信
            Receiver re = new Receiver(address, port);
            tRe = new Thread(re);
            tRe.start();
            disconnectButton.setEnabled(true);//改变按钮状态
            connectNormalButton.setEnabled(false);
        });

        disconnectButton.addActionListener((event) -> {
            tRe.interrupt();//请求终止接收线程!
            disconnectButton.setEnabled(false);
            connectNormalButton.setEnabled(true);
        });
        this.add(bPanel, BorderLayout.NORTH);
    }

/**
 * @author Administrator 服务接收线程
 */
class Receiver implements Runnable {
    private String host;
    private int port;

    public Receiver(String host, int port) {
        super();
        this.host = host;
        this.port = port;
    }

    @Override
    public void run() {
        //使用try资源块,自动关闭资源
        try (Socket s = new Socket();) {
            s.connect(new InetSocketAddress(host, port), 5000);
            //更新Swing控件
            EventQueue.invokeLater(() -> {
                textArea.append("连接成功:"
```

```java
            + s.getRemoteSocketAddress().toString() + "\n");
    });
    try(Scanner in=new Scanner(s.getInputStream(),"UTF-8");//对从套接字得到的输
                                                          //入输出流进行包装
            PrintWriter out = new PrintWriter(
                new OutputStreamWriter(s.getOutputStream(),
                    "UTF-8"), true);){//自动刷新缓冲区,注意字符编码要统一
        while (!Thread.currentThread().isInterrupted()){//当前线程没被请求终止
            if (in.hasNextLine()) {
                //如果没有输入,hasNextLine 方法则会阻塞,等待接下来的数据
                //先读出来,再交给其他线程
                String str = in.nextLine();
                EventQueue.invokeLater(() -> {
                    textArea.append(str + "\n");
                });
                //发送指定的命令
                out.println(sendComm.getSelectedItem().toString());
            }
        }
        out.flush();//强制更新缓冲区
        s.shutdownOutput();//关闭输出,处于半关闭
        //但仍然可以从输入中获取信息。
        while (in.hasNextLine()) {//hasNextLine 方法会阻塞,等待输入
            String str = in.nextLine();//将数据存到一个 String 对象,并交给
                                       //Swing 的更新线程曲更新
            //由于创建了一个线程并交给事件队列取更新 Swing 控件,
            //所以注意如果直接把 in.nextLine()写在 append 语句中,会导致数
            //据读写异常的出现,例如 in.nextLine 拿不到数据
            EventQueue.invokeLater(() -> {
                textArea.append(str + "\n");
            });
        }
    }
} catch (SocketTimeoutException e) {
    System.err.println("连接超时!!!!");
} catch (IOException e) {
    e.printStackTrace();
}
```

4. 程序分析与注意事项

此部分包括程序解释及常见问题。

① 在客户端中为了不影响 Swing 控件的更新，在单击【连接】按钮后启动一个线程用于和服务器连接，并进行消息的收发。本例中不允许一个客户端启动多个线程用于和服务器交互，因此注意对【连接】【断开】按钮状态的更新。

```
disconnectButton.setEnabled(true);connectNormalButton.setEnabled(false);
```

② 客户端中在接收到消息后，让事件队列对更新控件，注意语句的书写。

```
while (in.hasNextLine()) {//hasNextLine 方法会阻塞，等待输入
    String str = in.nextLine();//将数据存到一个 String 对象，并交给 Swing
                               //的更新线程曲更新
    //由于创建了一个线程并交给事件队列取更新 Swing 控件，
    //所以注意如果直接把 in.nextLine()写在 append 语句中，会导致数据
    //读写异常的出现，例如 in.nextLine 拿不到数据
    EventQueue.invokeLater(() -> {
        textArea.append(str + "\n");
    });
}
```

③ 半关闭状态，Socket 中的 shutdownOutput 方法将关闭输出，但仍然可以使用输入流，此时 Socket 处于半关闭状态。通信的另一方将得到相应的通知，从而得知对方不会再有输出。

④ 服务器端对每一个处理线程都会保存其执行结果，本例采用的是 List 类型，如：

```
List<Future<Integer>> serverThreadResult = new ArrayList<Future<Integer>>()
```

实践 8-2　利用 URL 访问 Web 服务器

1. 实践结果

本例的运行结果如图 8-4 所示。

2. 实践目的

本例介绍利用 URL 类和 URLConnection 类来访问 Web 网站，利用 GET 和 POST 的方法向 Web 服务器发送数据，并接收服务器返回的响应信息。

（1）URL 类

如果仅仅是获得资源的内容，则可以简单地利用 openStream 方法，例如：

```
InputStream inputStream = url.openStream()
Scanner in = new Scanner(inputStream, "UTF-8")
```

URL 类通常用于 http:、https:、ftp:、file:、jar:等模式。

（2）URLConnection 类

如果想从 Web 服务器获得更多的信息，使用 URLConnection 对象。例如 URLConnection

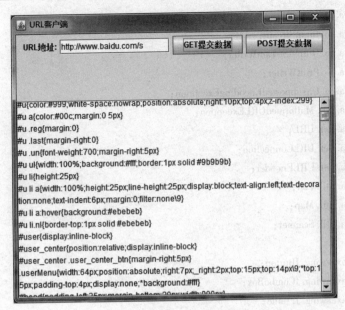

图 8-4 实践 8-2 的运行界面

conn = url. openConnection( ), 在获得 URLConnection 对象后, 可以进行如下操作。

getInputStream, getOutputStream: 获得输入输出流。

setDoOutput: 设置连接属性为可以执行写操作。

setRequestProperyt: 设置请求头。

getHeaderFilelds: 获得响应头字段的 Map 对象。

(3) 提交数据

可以利用 URLConnection 对象, 向 Web 服务器提交数据, 主要采用以下两种方法。

GET: 利用?和 & 构建查询字符串, 并创建 URL 对象, 例如 url = new URL(host+"?wd ="+URLEncoder. encode("华北科技学院","UTF-8")), 当参数多于一个时, 用 & 连接。之后利用 URL 类的 openStream 打开输入流, 获取服务器的响应即可, 如 InputStream inputStream = url. openStream( )。

POST: 利用 URLConnection 的 setDoOutput(true)方法建立一个用于输出的连接, 然后利用 getOutputStream 方法获得输出流, 最后向输出流中直接写入参数名和参数值即可。例如 out. print("wd" + "=" + URLEncoder. encode("华北科技学院","UTF-8")), 同样当参数多于一个时, 应当使用 "&" 进行连接。在利用输出流的 close 方法关闭输出流后, 可以调用 getInputStream 方法获得服务器的响应。

3. 程序源代码

程序的源代码如下:

```
package exp2;

import java. awt. BorderLayout;
import java. awt. Dimension;
import java. awt. EventQueue;
```

```java
import java.io.IOException;
import java.io.InputStream;
import java.io.OutputStreamWriter;
import java.io.PrintWriter;
import java.io.UnsupportedEncodingException;
import java.net.MalformedURLException;
import java.net.URL;
import java.net.URLConnection;
import java.net.URLEncoder;
import java.util.List;
import java.util.Map;
import java.util.Scanner;

import javax.swing.JButton;
import javax.swing.JComboBox;
import javax.swing.JFrame;
import javax.swing.JLabel;
import javax.swing.JOptionPane;
import javax.swing.JPanel;
import javax.swing.JScrollPane;
import javax.swing.JTextArea;
import javax.swing.JTextField;

public class Exp2 {
    public static void main(String[] args) {
        //TODO Auto-generated method stub
        EventQueue.invokeLater(() -> {
            MyClient client = new MyClient();
            client.setDefaultCloseOperation(JFrame.EXIT_ON_CLOSE);
            client.pack();
            client.setVisible(true);
        });
    }
}

class MyClient extends JFrame {
    JTextArea textArea = new JTextArea(30, 30);//为了在内部类里使用,所以放在这里
    JComboBox<String> sendComm = new JComboBox<String>();
    Thread tRe = null;

    public MyClient() {
```

```java
        super("URL 客户端");
        //TODO Auto-generated constructor stub
        this.setPreferredSize(new Dimension(500, 800));
        JPanel bPanel = new JPanel();
        bPanel.setPreferredSize(new Dimension(500, 100));
        bPanel.add(new JLabel("URL 地址:"));
        JTextField serverAddress = (JTextField) bPanel.add(new JTextField(
                "http://www.baidu.com/s", 15));//服务器地址

        JButton getMethodButton = new JButton("GET 提交数据");
        JButton postMethodButton = new JButton("POST 提交数据");

        this.add(new JScrollPane(textArea), BorderLayout.CENTER);
        textArea.setLineWrap(true);
        //设置自动换行,但是不会修改"\n"的个数

        bPanel.add(getMethodButton);
        bPanel.add(postMethodButton);
        //添加按钮监听器
        getMethodButton.addActionListener((event) -> {
            textArea.setText("GET 方式!!!!");
            String address = serverAddress.getText();

            //创建一个接收线程与服务器通信
            SendAndReceiver re = new SendAndReceiver(address, false);//Get
            tRe = new Thread(re);
            tRe.start();

        });
        postMethodButton.addActionListener((event) -> {
            textArea.setText("POST 方式!!!!");
            String address = serverAddress.getText();
            //创建一个接收线程与服务器通信
            SendAndReceiver re = new SendAndReceiver(address, true);//Post
            tRe = new Thread(re);
            tRe.start();

        });

        this.add(bPanel, BorderLayout.NORTH);
    }

/**
 * @author Administrator 服务线程
```

```java
    */
class SendAndReceiver implements Runnable {
    private String host;
    private boolean isPostMethod = false;//是否采用Post方式
    public SendAndReceiver(String host, boolean isPostMethod) {
        super();
        this.host = host;
        this.isPostMethod = isPostMethod;
    }

    @Override
    public void run() {
        URL url = null;
        if(this.isPostMethod) {//Post方式
            URLConnection conn;
            try {
                url = new URL(host);
                conn = url.openConnection();
                //请求头内容详见 https://en.wikipedia.org/wiki/List_of_HTTP_header_fields
                conn.setRequestProperty("Accept-Language", "en");//设置请求头
                conn.setRequestProperty("Referer", "Http://www.ncist.edu.cn");//设置请
                                                                              //求头
                conn.setDoOutput(true);//设置输出连接属性,默认情况下只能是输入
                try(PrintWriter out = new PrintWriter(new OutputStreamWriter(conn.getOutputStream(),"UTF-8"))) {
                    out.print("wd" + "="
                        + URLEncoder.encode("华北科技学院", "UTF-8"));
                                        //写入数据,多个数据时仍然要使用&
                    //对参数值使用URLEncoder进行编码
                }
                //获得响应头,并遍历输出到控制台
                Map<String,List<String>> headers = conn.getHeaderFields();
                for(Map.Entry<String,List<String>> ent:headers.entrySet()) {
                    String k = ent.getKey();
                    for(String v:ent.getValue()) {
                        System.out.print(k+":"+v);
                    }
                    System.out.println();
                }
                //获得来自服务器的输入数据
                try(Scanner in = new Scanner(conn.getInputStream(), "UTF-8")) {
                                        //对从连接得到输入流进行包装
```

```java
                int lineCount = 0;
                while (in.hasNextLine()) {//hasNextLine 方法会阻塞，等待输入
                    String str = in.nextLine();//将数据存到一个 String 对象，并交给
                                               //Swing 的更新线程去更新
                    //由于创建了一个线程并交给事件队列取更新 Swing 控件，
                    //所以注意如果直接把 in.nextLine()写在 append 语句中，会导
                    //致数据读写异常的出现，例如 in.nextLine 拿不到数据
                    EventQueue.invokeLater(() -> {
                        textArea.append(str + "\n");
                    });
                    lineCount++;
                }
                JOptionPane.showMessageDialog(null, "收到" + lineCount
                        + "行数据");
            }
        } catch (IOException e) {
            //TODO Auto-generated catch block
            e.printStackTrace();
        }
    
    } else {//GET 方式
        try {
            url = new URL(host+"?wd=" +URLEncoder.encode("华北科技学院",
"UTF-8"));//构造 Get 字符串，注意?和 & 的使用
            //对参数值使用 URLEncoder 进行编码
        } catch (MalformedURLException | UnsupportedEncodingException e1) {
            //TODO Auto-generated catch block
            e1.printStackTrace();
        }
        //使用 try 资源块，自动关闭资源
        try (InputStream inputStream = url.openStream()) {
            //更新 Swing 控件
            EventQueue.invokeLater(() -> {
                textArea.append("获得响应,来自" + host + "\n");
            });
            
            try(Scanner in=new Scanner(inputStream,"UTF-8")) {//对从套接字得到的输
                                                             //入输出流进行包装
                int lineCount = 0;
                while (in.hasNextLine()) {//hasNextLine 方法会阻塞，等待输入
                    String str = in.nextLine();//将数据存到一个 String 对象，并交给
                                               //Swing 的更新线程去更新
                    //由于创建了一个线程并交给事件队列取更新 Swing 控件，
```

```
                    //所以注意如果直接把 in.nextLine( )写在 append 语句中,会导
                    //致数据读写异常的出现,例如 in.nextLine 拿不到数据
                    EventQueue.invokeLater(( ) -> {
                        textArea.append(str + "\n");
                    });
                    lineCount++;
                }
                JOptionPane.showMessageDialog(null,"收到" + lineCount
                    + "行数据");
            }
        } catch (IOException e) {
            e.printStackTrace();
        }
    }
}
```

4. 程序分析与注意事项

① 为了防止用户界面出现假死情况,本程序创建单独的线程用于和 Web 服务器进行交互。

② 在设置参数值时,需要编码为 UTF-8 格式。通常利用 URLEncoder 类将每个字节都编码为%号后面紧跟一个两位十六进制数字的形式,如:

```
URLEncoder.encode("华北科技学院","UTF-8")
```

③ 在利用 getHeaderFields( )方法获得的响应头信息类型为 Map<String,List<String>>,遍历时需要注意。

④ 本例中的 TextArea 中内容较多,因此设置了自动换行。例如 textArea.setLineWrap(true);虽然设置了自动换行,但是并不会修改文本区域内"\n"的个数。

实践 8-3 利用 JavaMail 发送 E-mail

1. 实践结果

本例的运行结果如图 8-5 所示。

本程序正确执行的前提条件是已经注册好一个邮箱,同时邮箱的 SMTP 协议为打开状态。注意:很多邮箱需要进行单独设置,否则不接受 SMTP 协议。

2. 实践目的

本例介绍如何利用 Java Mail 发送 E-mail。首先需要从 https://github.com/javaee/javamail/releases/download/JAVAMAIL-1_6_2/javax.mail.jar 下载 JavaMail 的 jar 文件,其次将 javax.mail.jar 配置在项目的 BuildPath 中。编程的主要步骤如下:

图 8-5 收到的邮件截图

① 创建属性集,并设置属性值;
② 利用属性集创建 Session 对象;
③ 利用 Session 对象创建 MimeMessage 消息对象,并设置发送者、接收者、主题、消息文本等内容;
④ 利用 Transport 对象发送消息对象。

3. 程序源代码

程序的源代码如下:

```java
package exp3;
import java.util.Properties;

import javax.mail.Message;
import javax.mail.Session;
import javax.mail.Transport;
import javax.mail.internet.InternetAddress;
import javax.mail.internet.MimeMessage;
public class Exp3 {

    public static void main(String[] args) {
        //TODO Auto-generated method stub
        SimpleEmail se=new SimpleEmail();
        se.setFrom("jspjspjsp@sohu.com");
        se.setTo("jspjspjsp@sohu.com");
        se.setSubject("测试邮件");
        se.setContent("测试邮件正文");
        se.setHost("smtp.sohu.com");
        se.setUsername("jspjspjsp@sohu.com");
        se.setPassword("******");//应为实际密码字符串,这里简单用*代替
        if(se.sendMail()){
            System.out.println("邮件发送成功!");
        }else{
```

```java
            System.out.println("邮件发送失败");
        }
    }
}

class SimpleEmail {
    private String host;
    private String username;
    private String password;
    private String from;
    private String to;
    private String subject;
    private String content;

    public SimpleEmail() {
        super();
    }

    public boolean sendMail() {
        Properties prop = new Properties();//属性对象
        Session session;
        MimeMessage message;
        try {
            prop.setProperty("mail.host", this.host);
            prop.setProperty("mail.transport.protocol", "smtp");
            prop.setProperty("mail.smtp.auth", "true");
            //使用 JavaMail 发送邮件的 5 个步骤
            //1. 创建 session
            session = Session.getInstance(prop);
            //开启 Session 的 debug 模式,观察 E-mail 的运行状态
            session.setDebug(true);
            //2. 通过 session 得到 transport 对象
            Transport ts;
            ts = session.getTransport();
            //3. 使用邮箱的用户名和密码连接邮件服务器,发送邮件时,发件人需要提交邮箱的用户
            //名和密码给 SMTP 服务器,用户名和密码都通过验证之后才能够正常发送邮件给收件人
            ts.connect(this.host, this.getUsername(), this.getPassword());
            //4. 创建邮件
            //创建邮件对象
            message = new MimeMessage(session);
            //指明邮件的发件人
            message.setFrom(new InternetAddress(this.from));
            //指明邮件的收件人,现在发件人和收件人是一样的,那就是自己给自己发
```

```java
                message.setRecipient(Message.RecipientType.TO, new InternetAddress(
                    this.to));//收件人
                message.setRecipient(Message.RecipientType.CC, new InternetAddress(
                    this.to));//抄送
                message.setRecipient(Message.RecipientType.BCC, new InternetAddress(
                    this.to));//保密抄送
                //邮件的标题
                message.setSubject(this.subject);
                //邮件的文本内容
                this.content = "<h1 align='center'><font color='#ff0000'>"+this.content+"</font></h1>";
                message.setContent(this.content, "text/html;charset=UTF-8");
                //返回创建好的邮件对象
                //5. 发送邮件
                ts.sendMessage(message, message.getAllRecipients());
                ts.close();
                return true;
            } catch (Exception e) {
                e.printStackTrace();
                return false;
            }
        }

        public String getHost() {
            return host;
        }

        public void setHost(String host) {
            this.host = host;
        }

        public String getUsername() {
            return username;
        }

        public void setUsername(String username) {
            this.username = username;
        }

        public String getPassword() {
            return password;
        }
```

```java
    public void setPassword(String password) {
        this.password = password;
    }

    public String getFrom() {
        return from;
    }

    public void setFrom(String from) {
        this.from = from;
    }

    public String getTo() {
        return to;
    }

    public void setTo(String to) {
        this.to = to;
    }

    public String getSubject() {
        return subject;
    }

    public void setSubject(String subject) {
        this.subject = subject;
    }

    public String getContent() {
        return content;
    }

    public void setContent(String content) {
        this.content = content;
    }
}
```

4. 程序分析与注意事项

本程序较为简单，创建了 SimpleEmail 类，对 JavaMail 发送过程进行了简单的封装。为了更好地检测邮件状态，开启了 Session 的 debug 模式，这样就可以查看到程序发送 E-mail 的运行状态。

练习题

1. 编写图形界面的 Application 程序,其中包含两个标签 Label(一个用来显示提示文字,一个显示结果)、一个单行文本域 TextField,在 TextField 中输入主机名,将其 IP 地址显示在标签上。

2. 利用 Socket 类和 ServerSocket 类编写一个 C/S 程序,实现 C/S 通信。

3. 创建一个服务器,用它请求用户输入密码,然后打开一个文件,并将文件通过网络连接传送出去。创建一个同该服务器连接的客户,为其分配适当的密码,然后捕获和保存文件。在自己的机器上用 localhost 测试这两个程序。

练习题

1. 编写程序调用 Application 服务器，其中程序有两个标签 Label（一个用来显示客户文本是否正确，一个输入反本框 TextField，在 TextField 中输入上传名，添其IP地址最后结果反之。

2. 利用 Socket 类和 ServerSocket 类实现一个 C/S 结构，实现 CNS 地址

3. 因为一个服务器，能够为下面的服务用户端输入文本后，给各打开一个文件，并将它们通过网络发送给客户端，客户端不同段建显示读出，为方便调试验证，需后台数据和原有文件，客户端可列表上用 JoptPane 对话框显示每个段落。